India Studies in Business and Economics

The Indian economy is considered to be one of the fastest growing economies of the world with India amongst the most important G-20 economies. Ever since the Indian economy made its presence felt on the global platform, the research community is now even more interested in studying and analyzing what India has to offer. This series aims to bring forth the latest studies and research about India from the areas of economics, business, and management science. The titles featured in this series will present rigorous empirical research, often accompanied by policy recommendations, evoke and evaluate various aspects of the economy and the business and management landscape in India, with a special focus on India's relationship with the world in terms of business and trade.

More information about this series at http://www.springer.com/series/11234

N.S. Siddharthan · K. Narayanan
Editors

Technology

Corporate and Social Dimensions

Springer

Editors
N.S. Siddharthan
Madras School of Economics
Chennai, Tamil Nadu
India

K. Narayanan
Department of Humanities and Social Sciences
Indian Institute of Technology Bombay
Mumbai, Maharashtra
India

ISSN 2198-0012 ISSN 2198-0020 (electronic)
India Studies in Business and Economics
ISBN 978-981-10-1683-7 ISBN 978-981-10-1684-4 (eBook)
DOI 10.1007/978-981-10-1684-4

Library of Congress Control Number: 2016946953

Printed on acid-free paper

This Springer imprint is published by Springer Nature
The registered company is Springer Science+Business Media Singapore Pte Ltd.

Contents

About the Editors

N.S. Siddharthan is an Honorary Professor at the Madras School of Economics, Chennai, India, and Honorary Director of the Forum for Global Knowledge Sharing (a specialized, interdisciplinary global forum, with its registered office at the Institute for Human Development (IHD), New Delhi, India and having members and office bearers from various institutions and countries worldwide). His current research interests are technology and globalization, international economics, multinational corporations, and industrial organization. He has published several papers in journals such as *The Economic Journal (Wiley), Oxford Bulletin of Economics and Statistics (Wiley), The Journal of Development Studies (Routledge), Economics of Innovation and New Technology (Routledge), Applied Economics, Development and Change, Journal of Economic Behavior and Organization (Elsevier), Journal of Business. Venturing (Elsevier), Japan and the World Economy (Elsevier), Journal of International and Area Studies (Institute of International Affairs), International Business Review (Elsevier), The Developing Economies (Wiley), Weltwirtschaftliches Archiv (Springer), Transnational Corporations (UNCTAD), The Indian Economic Review (DSE), The Indian Economic Journal (Indian Economic Association), and Sankhya (Springer).* He has also served as a guest editor for special issues of the journals *Science, Technology and Society* (Sage) and *Innovation and Development* (Routledge). He has also published books with publishers such as Springer, Routledge, Oxford University Press, Macmillan, Allied, Academic Foundation, and New Age International Publishers.

Dr. K. Narayanan received his Ph.D. in Economics from the Delhi School of Economics, University of Delhi, India, and pursued postdoctoral research at the Institute of Advanced Studies United Nations University, Tokyo, Japan, during 2000–2001. During his research studies in India, he was a recipient of the ICSSR fellowship at the Institute of Economic Growth, New Delhi, India. His research interests span the areas of industrial economics, international business, socioeconomic empowerment through ICT, environmental economics, economic impacts of climate change, and development economics. He has a number of publications in the field of industrial competitiveness, technology transfer, information and communication technology, international trade, and the socioeconomic impacts of climate change. The research journals in which he has published include *Research Policy (Elsevier), Journal of Regional Studies, Technovation (Elsevier), Oxford Development Studies (Taylor and Francis), International Journal of Energy Economics and Policy (EconJournals), and Economic and Political Weekly*. Two of his recent publications include edited books on (i) Indian and Chinese Enterprises: Global Trade, Technology, and Investment Regimes and (ii) Human Capital and Development: The Indian Experience [both of which were jointly edited with N.S. Siddharthan published by Routledge and Springer, respectively]. He also guest-edited a special issue of the Indian Association of Social Science Institution's quarterly on the theme "Human Capital and Development," as well as a special issue of the Sage published international journal, *Science, Technology and Society* on the theme "Agglomeration, technology clusters and networks." He is actively engaged in a Web-based research group, Forum for Global Knowledge Sharing, which brings together scientists, technologists, and economists. He is currently the Institute Chair Professor at the Department of Humanities and Social Sciences, Indian Institute of Technology Bombay, Mumbai, India.

Introduction

N.S. Siddharthan and K. Narayanan

1 Technological and Institutional Background

The use of technological innovation to foster industrial development is not a new phenomenon. Few would disagree that technology underpins the unprecedented levels of prosperity enjoyed by developed and some of the fast growing developing countries. Industrial sector in India has been subjected to substantial policy changes over the last three decades. The policy changes were in two doses and took the form of partial de-regulations introduced in 1985 and liberalisation measures launched since 1991. The pre 1985 regime could be described as an era of strict controls and regulations. Liberalisation of economic policies and the outward orientation introduced since 1991 has brought about a dramatic change in Indian industries. These policy measures considerably transformed the environment in which the firms had been operating. As a consequence, most industries witnessed the entry of new firms involving foreign equity and adoption of strategies by the already existing firms to introduce technological change and improve their performance. The new players brought in modern engineering, efficient processes and effective shop-floor layouts. The new manufacturing strategies include breaking up of the plant into modules and cells; reduce the complexity of purchasing logistics, reduction of inventories and product complexity, and creation of simpler processes by encouraging flexibility and teamwork. These firms also make extensive use of CAD/CAM in their plants. Moreover, the materials used in most industries have also undergone a change from traditional steel and cast iron to aluminium and

N.S. Siddharthan (✉)
Madras School of Economics, Chennai, Tamil Nadu, India
e-mail: nssiddharthan@gmail.com

K. Narayanan
Department of Humanities and Social Sciences, Indian Institute of Technology Bombay, Mumbai, Maharashtra, India
e-mail: knn.iitb@gmail.com

N.S. Siddharthan and K. Narayanan (eds.), *Technology*,
India Studies in Business and Economics,
DOI 10.1007/978-981-10-1684-4_1

thermoplastics. Some of the existing firms have oriented their systems by replacing the batch system by work flow, organising the production in product modules and by keeping the product-mix flexible in order to save time, reduce cost and increase quality. The new joint ventures, it appears, are becoming catalysts to activate the capabilities of the existing plants in areas of cost control and product development. As a result, the policy changes to introduce market-induced efficiency have had far reaching implications in the form of technology acquisition, growth in output and exports.

Further, It is widely recognized that one of the vital determinants of competitiveness in developing countries is the skill of the workforce at all levels. The effective use of technologies requires skills, and the move from simple to complex technologies requires more, better and more diverse skills. Further, new technologies often call for entirely different skills, both for direct production and services and for the organization of production and for managing knowledge networks. For example, as the momentum surrounding e-commerce increases with more and more people turning to World Wide Web to conduct business transactions, the demand for Web professionals is reaching a feverish pitch. The need for increased skills rises with the level of development, but even the least developed countries have to improve their human capital base if they are to grow and prosper. Traditional models of competition based on low costs and prices are being replaced by competition driven by quality, flexibility, design, reliability and networking. This change is not only in markets for advanced manufacturers but also in day-to-day use consumer goods like clothing, footwear and food products.

The use of new technologies, especially information-based technologies, calls for more, better and newer kinds of skills. The reason for this stems largely from both technological factors, as well as organizational factors. These skills are subject to constant change. Consequently, the education and training system has to upgrade skills constantly in line with emerging needs. International Labour Organisation observed that the demand for professionals and technicians has increased in all countries, as their analytical, cognitive and behavioural skills equip them better to adapt to more sophisticated technology. In addition, the increased importance of networking between firms (and between firms and technology institutions) for competitiveness requires specific "communicative" skills. Earlier studies have shown how these skills are enabling skilled workers in the UK to move into knowledge-intensive sectors more readily than worker without such skills. It is widely believed that skilled and experienced individuals can provide an economy with greater technological capabilities for fostering growth in output and social welfare.

In a world of fast changing technology, however, a base of formal education is necessary but not sufficient for efficiently using technologies. Technology-specific skills and learning are as much important as the formal education is. Educational Institutions in both developed and developing countries tend to cater to this requirement. Increasing the skill content of the potential work-force open up a wide range of avenues to them. Better capabilities, especially in micro-electronic applications and use of information technology, for job seeking is expected to bring about a change in the socio-economic structure of the society. This is especially when such skill formation efforts, in developing countries, are directed to socially and economically underprivileged.

2 Technology Efforts of Indian Corporate Sector

Innovation is considered as one of the most significant drivers of market competition in the context of rapid changes in the pattern of production, and nature and extent of competition. With production becoming more and more knowledge-oriented across a wide range of industries and the process of liberalization and globalization leading to increase in market competition, emergence of innovation-based competition is imperative. While the developed country firms make significant in-house R&D efforts, technological progress in the developing countries takes place mainly through spillovers from trade, foreign direct investment, technology licensing, joint ventures, mergers, acquisitions and various other alliances.

The long-term competitive ability of Indian firms depends on production efficiency. Production efficiency, in turn, is dependent on ability to develop, import and adapt new technologies among other factors. India has made significant progress in various spheres of science and technology over the years, and hence has a strong network of S&T institutions, trained manpower and an innovative knowledge base. Technology development is critical to a country's efforts in improving productivity, efficiency and competitiveness of its industrial sector. Factor cost advantages are being replaced by technology related factors such as zero-defect product quality and international certification of firms' quality assurance systems (e.g., ISO 9000) in determining international competitiveness. Indian firms present a full spectrum of technological capabilities—while there are few firms close to the international frontier in terms of product design capability and process technology, technological capabilities of most players are extremely limited due to growing technological obsolescence, inferior quality, limited range and high costs. Most of the Indian manufacturing firms appear to be stuck at the basic or intermediate level of technological capabilities. Though Indian manufacturing industry has mastered standard techniques it has remained dependent for highly expensive and complicated technologies.

The Indian government has put in significant effort in the last 50 years to develop a vibrant scientific and technical infrastructure in the country. The combination of state-of-the-art infrastructure and highly qualified manpower ensures that India is poised to be the next Global R&D hub. This is increasingly being observed in Industry as large MNEs including GE, Microsoft, Bell Labs etc. have opened there R&D Centres in India—a first outside US for most of these companies. It must be noted that R&D Centres are not only restricted to IT and ITES sectors, but also they are present in all other sectors such as the Pharma, Chemical,

Automobile etc. India may be behind China in manufacturing, but it has taken an early lead in attracting R&D investments. India is emerging as a serious contender as a base for new offshore R&D centres.

The Government of India allocates a budget for scientific and technological (S&T) activity under an R&D fund. The allocation has increased from USD 18 million in 1950–51 to USD 2800 millions in 2000–01. In comparison, the share of industry in R&D has become of the order of $547 million, about 20 % of the government's contribution. The percentage share of major scientific agencies in total S&T expenditure is approximately 70 %. In the total S&T expenditure by the government, the share of non-scientific ministries has been approximately 30 % combined for all sectors, including agriculture, rural development, energy, industry and minerals, transport, communication and others. The total expenditure on R&D, including from industry, is about 0.8 % of GNP for the past several years. Compared to most advanced countries, which spent between 4 and 6 % on R&D, this proportion is quite low.

Table 1 presents the expenditure by the private industrial sector in R&D in India, with number of R&D units. We can observe that the number of R&D units vary from a single unit to 161 units in some sectors. There are around 1108 R&D units as given by the data of Government of India as of 2005–06. The R&D expenditure

Table 1 Expenditure on R&D by industry groups for private sector

Sl. No.	Industry group	No. of R&D units	R&D expenditure (Rs. Crores) 2004–05	R&D expenditure (Rs. Crores) 2005–06
1	Boilers and steam generating plants	1	290.71	431.25
2	Timber products	2	23.92	17.76
3	Consultancy services	2	25.18	26.99
4	Glass	3	204.68	283.42
5	Photographic raw film and paper	3	463.87	592.88
6	Prime movers	3	2548.51	4036.97
7	Vegetable oil and vanaspathi	3	3199.18	3509.92
8	Fermentation industries	4	653.20	511.75
9	Leather, leather goods and pickers	4	655.72	769.16
10	Glue and gelatin	4	147.64	203.33
11	Fertilizers	7	736.90	1183.62
12	Earth moving machinery	7	1239.97	1355.90
13	Defence industries	7	1580.43	1567.44
14	Cement and gypsum products	8	1392.04	1831.80
15	Agricultural machinery	8	5685.55	4863.55
16	Fuels	9	986.89	1085.47

(continued)

Table 1 (continued)

Sl. No.	Industry group	No. of R&D units	R&D expenditure (Rs. Crores) 2004–05	R&D expenditure (Rs. Crores) 2005–06
17	Dye-stuffs	9	598.41	593.68
18	Scientific instruments	9	439.25	486.30
19	Soaps, cosmetics and toilet preparations	10	17,387.77	13,722.41
20	Commercial offices, household equipment	11	2629.91	2553.22
21	Machine tools	11	467.26	712.30
22	Medical and surgical appliances	11	1040.61	1203.52
23	Ceramics	16	906.38	991.20
24	Papers and pulps	18	4955.64	6721.70
25	Sugar	19	1482.87	1575.73
26	Information technology	20	22,123.63	30,663.30
27	Food processing industries	24	3563.26	4086.68
28	Rubber goods	27	2182.40	2717.77
29	Industrial machinery	29	4096.53	3970.15
30	Telecommunications	30	8181.37	9867.14
31	Textiles (dyed, printed, processed)	33	2770.30	2513.24
32	Industrial instruments	33	5133.35	6869.49
33	Miscellaneous industries	39	3872.44	4499.20
34	Metallurgical industries	50	13,182.73	14,287.15
35	Misc. mechanical engineering industries	55	11,086.27	13,730.69
36	Transportation	60	86,279.81	104,719.74
37	Biotechnology	70	18,750.86	27,773.66
38	Electricals and electronics equipment	132	29,042.00	37,540.45
39	Drugs and pharmaceuticals	156	22,3712.19	282,686.34
40	Chemicals (other than fertilizers)	161	23,862.35	30,086.52
	Total	1108	507,581.98	626,842.79

Source Government of India

shows an increasing trend as depicted in the table for all most all the sectors of the economy.

Table 2 shows the two-digit classification of industrial (represents both the private and public) sector with number of R&D units and R&D expenses from 2007 to 2009. We can observe that the number of R&D units widely differ in the two-digit NIC classification. However, the R&D spending for most of the

Table 2 R&D expenditure of Indian Corporate Sector (Rs. in Crores) and number of R&D units

Sl. No.	Industry group	NIC code	No. of R&D units	2007–08 (Rs. in Crores)	2008–09 (Rs. in Crores)	2009–10 (Rs. in Crores)
1	Chemicals (other than fertilizers)	20	497	9.87	11.54	13.59
2	Commercial offices, household equipments	26	1105	41.08	40.43	42.22
3	Drugs and pharmaceuticals	21	672	15.76	17.45	19.46
4	Electricals and electronics equipment	27	813	12.82	12.98	12.2
5	Industrial machinery	28	3375	80.9	81.21	92.7
6	Leather, leather goods and pickers	15	693	15.22	13.51	15.17
7	Metallurgical industries	25	1646	24.46	23.77	22.12
8	Other manufacturing (gems and jewellery)	32	118	1.23	1.29	0.97
9	Papers and pulps	17	83	1.35	1.74	2.11
10	Rubber goods	22	83	1.38	1.81	1.58
11	Textiles (dyed, printed, processed)	13	1436	28.08	28.65	29.47
12	Timber products	16	63	0.64	0.47	0.77
	Total		10,584	232.79	234.85	252.35

Source Government of India

sub-industries have increased from 2007 to 2009. The data as shown in Table 3 shows that firms in low turnover ranges from below Rs. 10 million to Rs. 50–100 million are showing a drastic dip in number of firms spending on any R&D even though by and large the total number of firms in such turnover range is increasing rapidly.

The policy initiatives seem to have made firms in India invest more in R&D; the in-house R&D intensity shows an increasing trend in the post-reform era and has increased from less than 1 % in 2000 to about 4 % of sales in 2010–11. Although the Indian firms still rely largely on foreign technology, more specifically on imports of capital goods, increase in R&D expenditure at an average rate of around 4 % per annum seems to be encouraging for a developing country like India. However, the foreign technology purchase intensity has fluctuated during the post-reform period. The reliance on foreign technology again increased during the phase of high economic growth till the beginning of the slowdown phase. Overall, disembodied technology purchase has declined and in-house R&D has shown an increasing trend (Table 4).

Table 3 Number of firms spending on R&D

Turnover (Rs. millions)	1996		2000		2010		2013	
	No. of firms	Firms that spend on R&D	No. of firms	Firms that spend on R&D	No. of firms	Firms that spend on R&D	No. of firms	Firms that spend on R&D
<10	484	6	797	1	2031	1	2129	0
10–20	258	7	335	6	586	2	474	0
20–30	207	7	260	5	407	7	318	3
30–40	197	8	187	5	315	4	240	3
40–50	173	12	175	6	264	5	179	2
50–100	583	44	683	39	895	29	698	16
100–200	729	64	811	62	1082	60	638	28
200–300	505	89	623	66	738	72	619	33
300–400	287	43	433	61	508	54	432	22
400–500	250	57	305	52	374	38	328	23
500–1000	625	176	845	151	1102	163	1067	112
1000–2000	471	197	621	157	949	182	971	144
2000–3000	180	94	276	92	389	94	462	90
3000–4000	72	42	131	59	232	72	284	66
4000–5000	54	26	74	20	156	54	202	52
5000–10,000	112	71	162	80	326	133	543	173
10,000–50,000	89	52	143	67	277	116	566	208
50,000>	17	14	29	22	70	38	144	78

Table 4 Trends in technological strategies in Indian Corporate Sector

Year	In-house R&D intensity	Domestic technology purchase intensity	Foreign technology purchase intensity		
			Royalty	Capital imports	Total
2000	0.06	0.08	0.32	2.35	2.67
2001	0.08	0.22	0.32	1.12	1.44
2002	0.08	0.23	0.20	0.75	0.95
2003	0.06	0.24	0.19	0.68	0.88
2004	0.12	0.30	0.17	0.94	1.11
2005	0.14	0.30	0.18	0.97	1.15
2006	0.19	0.29	0.20	1.11	1.31
2007	0.20	0.28	0.21	1.44	1.65
2008	0.21	0.30	0.31	1.53	1.85
2009	0.31	0.28	0.26	2.08	2.34
2010	0.32	0.27	0.29	2.33	2.62
2012	0.35	0.32	0.24	1.73	1.97
2013	0.37	0.34	0.20	1.56	1.76

Source CMIE Prowess. The intensities are calculated by dividing the values by net sales

3 Social Dimension of Technology: A Brief Overview

As the industrial sector grows more complex and sophisticated, the challenge of providing better and more appropriate human capital becomes more important. In the process, therefore, relevant institutions develop. However, given the complexity of the information involved, the long-term nature of skill investment and the inherent uncertainties and externalities, there can be widespread market failures in human capital formation. As a result, markets can fail to provide properly an economy's skill needs and to keep with its changing profile. There is a clear need for policy support from the government. Evidence show that, surprisingly, it is the richest countries, with the most developed skill bases, that worry most about human capital problems. Human development despite faltering economic development has been taking place in India, how much ever gradually, in many states. The state of Kerala in India appears to be an exception to this, especially in achieving human development. While the formal Government sector continues to play an important role, it is the growth of private participation in providing affordable computer education, especially to the socially and economically under-privileged that appear to have provided the impetus for growth in qualified professionals. There is a broad consensus of appreciation on the role of Non-Governmental Organizations, Private Charitable Trusts and Educational Investment of Business Houses in fostering the growth of supply in computer educated and qualified personnel. This consensus emerged from the experience of market failure situation in providing quality educational services in developing countries.

In the literature on education and training system in developing countries, it is well recognized that market failure can take place due to (a) information gaps and uncertainty where the individuals may not know of the future value of investments in education and training or of particular skills, and they may not know what skills are needed in future, (b) even if the individuals can forecast the probability of getting returns on skill investments, they may prefer more certain short-term returns to available jobs, (c) individuals may not be able to finance their learning costs and foregone earning, especially with their inability to afford two square meals a day and (d) high costs of educational services provided, especially in the case of private institutions and (e) danger of bureaucratic and rigid management especially in a publicly funded training institution. As a result, government and corporate sector have a large role to play in imparting training and knowledge, especially for socially and economically underprivileged.

Moreover, decentralized approach to development has gained considerable acceptance in policy circles in the Indian development process during the past three decades. Most of the studies on the impact of decentralized development plans have emphasized the need for community participation to undertake tasks of identifying appropriate technology, adopting cost-efficient methods in production processes, and prioritize the technological and other policy support. However, even if the communities are well-motivated, dynamic and organized, inadequacy of material resources, incomplete information and access to appropriate technology may present severe limitations for achieving productive efficiency and higher socio-economic benefits. Institutional support to overcome this limitation, more so in providing technological and marketing support for instance, can go a long way to help various community and rural industries and different self-help groups towards achieving economic independence and empowerment. How best could technologies be used to achieve these objectives needs to be explored very clearly.

It is in this context that the collected papers in this volume are presented to cover both the corporate as well as social dimension of technology.

4 About the Volume

The book deals with two of the vital dimensions of technological activity; the corporate as well as the social dimension. The papers on corporate dimensions deal with the impact of patents, determinants of innovative activities, differential behaviour of multinationals, industrial groups and other firms with regard to innovations and technology. The papers on social dimensions mainly deal with the role of technology in reducing inequality. Most of the papers use econometric techniques and other statistical methods. Some of them are also based on primary data.

Unlike studies that mainly concentrated on R&D expenditures, the papers included in this volume on corporate dimensions concentrate on R&D output, namely, patents. The studies emphasise the importance of innovations and in particular patents, and the quality of human capital in influencing productivity across Indian states, significance of patenting in determining efficiency of firms, role of business groups in promoting innovations, differences in the technological characteristics of multinational and domestic firms, and mergers and acquisitions in promoting research and development.

The papers on social dimensions analyse the role of innovative activities in influencing employment, impact of technology on rural poverty, socio economic characteristics that influence the mobile phone ownerships, factors influencing the use of information and communications technologies in education institutions, and influence of Synchronous Technologies in reducing access to teaching programmes.

The studies show that the Indian States that invested in human capital and technology resulting in more number of patents experienced higher labour productivity in manufacturing. Studies prefer the use of patents rather than R&D to analyse the impact of innovations on technical efficiency and productivity. Furthermore, the results of the studies indicate that Business Group affiliates show more R&D intensity and obtained more patents compared to stand-alone firms. In capital intensive industries like the Indian machinery sector multinational enterprises showed higher efficiency in resources use but preferred to import technology rather than performing in-house R&D. To enhance their innovation capabilities several firms have been acquiring R&D intensive firms. Thus with regard to in-house R&D, the decision of the enterprise is not confined to make or buy technology. There is one more option, instead of buying technology, buy a technology intensive firm.

With regard to the use of technology for reducing inequalities, studies included in the volume clearly establish a positive relationship between R&D spending and employment. Adoption of agriculture related technologies have a significant impact on the reduction of rural poverty and increase in rural consumption expenditures. The usefulness of mobile phones is by now well recognised. However, the study included in this volume reveals that human development index and education levels of the individuals are still the main determinants of its ownership in India. Likewise, the importance of information and communication technologies in the field of higher education is also well known. Despite this, its penetration in the Indian higher education institutions is not high. The studies included in this volume discuss the determinants of penetration of ICT.

5 A Brief Tour of Chapters

In presenting the corporate dimensions, we begin with inter-state differences and then move on to inter-firm (enterprises) studies. The paper by Savita Bhat analyses the determinants of inter-state differences in labour productivity and the role of human skills and innovative factors in influencing productivity. Using panel data for three years (2011–13) for 18 large Indian States, the paper empirically examines the role of state-level innovative efforts, represented in terms of the state-level patent applications filed, in determining the differences in industrial labour productivity. Her results indicate that after controlling for human capital and fixed capital, differences in innovative efforts do explain inter-state differences in industrial labour productivity. In addition human capital indicators representing health and education characteristics of the states like infant mortality rate and percentage of schools with computers also emerge as important determinants.

The next paper by Abrammal and Sharma explores the impact of patenting on the technical efficiency of enterprises. Their sample consists of 489 high and medium technology firms for the period 2000–2010. The study discusses efficiency improvement of firms as a function of its investment in innovation. They considered three variables to denote innovations—R&D, patents and technology purchases. They found patents and technology imports important in explaining technical efficiency and not R&D. In this context it is worth noting that technology purchases against technical fees and payments, by and large, involve fees paid to use patents registered by others. Furthermore, R&D is an input while patents are outputs. Firms could incur R&D expenditures for tax and other purposes that need not result in the creation of new products and processes. On the other hand patents are more reliable. These two studies are important as they are among very few studies using Indian data to rigorously test the role of patents.

The next two papers deal with two other important factors that could have a bearing on innovative activities. They deal with the influence of ownership of the enterprises on their innovative activities. The first paper by Komera, Lukose, and Sasidharan deals with the impact of affiliation to Industrial houses and the next by Kesari with the affiliation to multinational enterprises. With regard to the Business Groups, the results are mixed. The results indicate that Business Group affiliates have a higher R&D intensity and better record of patents compared to stand-alone firms. However, in their regression equations, after controlling for firm specific characteristics, business group affiliation seems to hinder innovation in technology intensive sectors. Kesari's paper empirically examines the differential behaviour of multinational enterprises and domestic firms in the machinery sector. His results show that foreign direct investments had resulted in a better resource use and technical efficiency of multinational affiliates, and they are also more export intensive. However, multinationals don't spend more on in-house R&D and prefer to import technology for improving efficiency.

The paper by Vidhisha and Narayanan argue that Mergers and acquisitions (M&A) has become a means for firms to acquire, absorb and exploit the knowledge assets of target firm. Their study investigates the relationship between M&A and innovation activities of pharmaceutical firms (post-M&A R&D intensity) for three post acquisition years and for average of three years. The findings of the paper suggest that acquisitions per se have a mixed impact on R&D intensity of firms in immediate post-acquisition years. While technology imports through embodied capital goods boost acquiring firms' R&D intensity, imports of disembodied technology lowers the need for R&D expenditure. Relatedness of target and acquiring firms and cross-border M&A, however, enhances R&D intensity in post-M&A period. This could imply that mergers and acquisitions of firms, if carried out in similar line of business, could boost innovative efforts in this industry in India.

There are not many studies dealing with the impact of R&D on employment, and even fewer at firm level. The study by Mitra fills this gap in literature. For a sample of Indian firms across 11 industries the study estimates the firm specific time variant total factor productivity growth and technical efficiency and assesses the impact of R&D on these performance indicators. The study finds a positive influence of R&D intensity on employment in high tech industries like engineering, households and personal products, drugs and pharmaceuticals. For other industries it was not significant. Furthermore, both exports and imports were positively associated with employment indicating complementary relationships between exports and imports.

The paper by Sahu and Das uses data from a primary survey to analyse the impact of the adoption of agriculture related technologies on rural consumption expenditures and rural poverty. This issue is of importance as there is a view that argues that the adoption of modern technology would increase rural poverty and favour only some rich farmers. This study shows that such fears are unfounded. Technology has a positive impact on rural per capita consumption levels and reduces poverty. They further found that the gain in consumption expenditures and reduction in poverty was highest in the lowest farm size quartile.

Paul and Murti examine scope and magnitude of the expanding telecommunication, and linkages to advances in development. They plot the linkages between demographic and economic factors and mobile phone penetration in India. The study shows that socio-economic characteristics mainly determine mobile phone penetration. This result is unexpected as mobile phones are relatively inexpensive and does not require much skill to operate. Most persons who can recognise numbers and alphabets can operate. It is not like a computer or email. Despite this the study finds that penetration levels are more among the educated and those enjoying a higher human development index.

The favourable impact of the information and communications technology (ICT) on education is by now well recognised. Despite the obvious advantages its penetration in several Indian higher education institutions has been low. It is important to find out the causes. The study by Lal and Paul aims at identifying major impediments that have resulted in slow penetration of ICTs in higher education institutions in rural India. The study is based on a primary survey of graduate and post graduate students. By and large, students from rural areas blame two factors for low penetration: inadequate infrastructure and lack of faculty motivation and unwillingness to put in extra effort.

The paper by Kalpana Kannan and Narayanan describes how technologies can be used to reduce the unequal access in teacher training in the context of an emerging economy like India. They collected data from the large-scale synchronous distance mode teacher training workshops for engineering college teachers, undertaken by the Indian Institute of Technology Bombay (IITB), under the National Mission on Education through ICT from the year 2009 to 2013. Unequal access and inability to improve outreach have remained as major constraints in conventional approach to teacher training, Statistical analysis of data about the beneficiaries and their feedback on training indicate that outreach of these workshops was good in terms of the number of participants, number of cities/towns, and number of states and colleges covered. Also, inclusivity in terms of gender, age group, location and educational qualification were well achieved, which usually is very difficult to achieve in a conventional face-to-face training. They compared the gender, age group, educational qualification and location of participating teachers in three workshops in the area of Computer science and engineering (CS) and three workshops in the area of Mechanical engineering (ME). The analysis indicate a clear difference in terms of gender, age and educational qualifications in CS and ME workshops. The comparison of results of Database Management System and Thermodynamics workshops after scaling up the number of participants show that there was not much change in the demography of the participants due to scale up. However, according to them, as the number of remote centres increased, the number of participants from rural colleges also increased, thus, reducing the gap between the urban and rural college teachers. Thus, it is evident that by using technology one can reduce the unequal access in teacher training and teaching material.

In short the social dimensions deal with the impact of technology on employment, rural consumption and poverty. They also discuss the determinants of technology penetration relating to connectivity and mobile phones, and information and communications technology and higher education and teacher training programmes.

Part I
Corporate Dimensions

Determinants of Inter-state Differences in Industrial Labour Productivity: Exploring the Role of Innovative Efforts

Savita Bhat

1 Introduction

Economic growth and development is a well-researched topic in the area of economics. Theories have been proposed that have incorporated technology or innovation as one of the factors that can stimulate economic growth. According to Verspagen (2005), during 1980s and 1990s, there emerged two dominant approaches to the analysis of the relationship between technology and growth, namely, neoclassical (e.g. Solow 1956; Grossman and Helpman 1994) and new-Schumpeterian or evolutionary theory (e.g. Nelson and Winter 1982). In a recent work Howitt (2000) proposes a multicountry Schumpeterian growth model where productivity differences due to research and development (R&D) explain cross-country income differences.

These growth models have led to several empirical studies in the context of both developed and developing countries that have tried to understand the factors that affect economic growth (Akçomak and ter Weel 2009; Banerjee and Roy 2014). Many of these studies have specially considered the issue of convergence in economic growth (see Verspagen 2005 for details). In the context of India, recent empirical studies on economic growth and development have invariably adopted variants of the standard growth model over time series data to study factors leading to convergence or divergence amongst states or districts (see Purohit 2008 for details; Agarwalla and Pangotra 2011). In a recent article, Bhat and Siddharthan (2013) observe that there is skill bias in the current technological revolution where employment and productivity grow faster in states that are better endowed in terms of human capital. However, in the presence of inter-state migration in India (Chandrasekhar and Sharma 2014), there is a possibility that high skill labour move from one state to another and contribute to income generation in latter states.

S. Bhat (✉)
School of Management, National Institute of Technology Karnataka (NITK), Surathkal, Mangalore 575025, Karnataka, India
e-mail: savitapbhat@gmail.com

N.S. Siddharthan and K. Narayanan (eds.), *Technology*,
India Studies in Business and Economics,
DOI 10.1007/978-981-10-1684-4_2

In the light of the above facts, this paper attempts to investigate the factors that explain inter-state differences in industrial labour productivity for India based on more recent panel data. In particular, the paper tries to understand the role of innovative efforts, defined in terms of patent applications filed, in explaining the inter-state differences in the industrial output per worker in India.[1]

2 Literature Review

The well-established Cobb-Douglas production function, output can be considered as a function of labour and capital. Several researchers have tried to extend this basic production function model to explore and emphasize the role of other factors in determining economic growth. In the Solow-Swan model technological progress is the additional factor and in Mankiw-Romer-Weil model human capital is the additional factor (for details see Oosterbaan et al. 2000). The following sub-sections review the literature that explains how innovation and human capital can affect output in an economy.

2.1 Innovation and Productivity

Economic theories have given importance to the role of innovation in explaining economic growth and development (Grossman and Helpman 1994). The proponents of innovation systems approach attempt to understand how innovation affects economic development at national level (Lundvall 1992; Nelson 1993; Freeman 1995) and regional level (Cooke et al. 1997). They emphasize the role of institutions and interactions in promoting innovation led development. Regional innovation networks lead to economic development as they make firms more competitive to provide higher wages and pay more taxes (Rutten and Boekema 2007).

Empirical studies in the context of developed countries have suggested a strong influence of technology development or innovation on economic growth and development (Ogburn and Allen 1959; Akçomak and ter Weel 2009). In a recent empirical study, Lin (2015) found that firms with headquarters near a knowledge centre are likely to enjoy better stock market valuation and growth. Whether a similar relationship holds true between innovative efforts and industrial labour productivity at state level needs to be explored in the context of India.

[1]The paper follows Bhat and Siddharthan (2013, p. 14) to propose that labour productivity is a function of choice of technology, that is, capital-labour ratio and other variables.

2.2 *Human Capital and Productivity*

Human capital is considered to be one of the important factors for economic growth and development (Barro 1991, 2001) as it can directly increase labour efficiency (Banerjee and Roy 2014). Education and health are considered to be important aspects of human capital formation (Maitra and Mukhopadhyay 2012). While education can provide returns that could be both monetary and non-monetary (Oketch 2006), health indicators like infant mortality rate can capture the general level of health, nutrition and wellbeing of the population (Bhat and Siddharthan 2013). Higher education indicators can also be used as a proxy for quality of labour or skill formation (Maiti and Mitra 2010).

Empirical studies in the context of developing countries have found a positive relationship between education and income inequality (Gregorio and Lee 2002; Paweenawat and McNown 2014). Studies have also found health expenditure (Maitra and Mukhopadhyay 2012) and health capital (Knowles and Owen 1995) to have positive effect on income.

Some recent studies have also explored the role of introduction of information and communication technology (ICT) in schools on residential adoption (Mo et al. 2013) and its spill-over effects outside schools (Tengtrakul and Peha 2013). The studies did find favourable effects of ICT in schools on skills of the students and on spill-over effects outside the school in terms of adults in the household using ICT.

3 Data, Sample and Descriptive Statistics

For the purpose of analysis secondary data has been collected from various publicly available government documents. Data for three consecutive financial years ending in 2011, 2012 and 2013 has been collected for 18 bigger states of India (excluding Delhi).[2] The details on the sources of data are in Appendix I. Suitable measures have been constructed to capture industrial labour productivity, industrial fixed capital to labour ratio, innovative efforts and human development. Table 1 gives the definitions of the variables. Innovative efforts are measured as number of ordinary patents filed at state level. Human capital is considered in terms of indicators of health and education. The health indicator considered is infant mortality rate (IMR).

Quality of education is captured using three different variables. One is the gross enrolment ratio in upper primary school (GER_UPP). This measure gives a general idea about the importance given to education beyond basic primary level by the people in the state. The second measure tries to measure the quality of elementary

[2]The Bigger States are identified as per the classification given in Sample Registration System (SRS) Bulletin September 2013. The Bigger States considered are Andhra Pradesh, Assam, Bihar, Chhattisgarh, Gujarat, Haryana, Jammu and Kashmir, Jharkhand, Karnataka, Kerala, Madhya Pradesh, Maharashtra, Odisha, Punjab, Rajasthan, Tamil Nadu, Uttar Pradesh and West Bengal. In this study, Delhi is not considered although it is classified under Bigger States in SRS Bulletin 2013.

Table 1 Definitions of the variables

Sl.	Variables	Symbol	Definition
1.	Industrial labour productivity	ILP	Ratio of inflation adjusted value of industrial output in Rs. Lakhs to number of industrial workers in a given year in the state
2.	Industrial fixed capital labour ratio	IFCL	Ratio of inflation adjusted fixed capital in Rs. Lakhs to number of industrial workers in a given year in the state
3.	Innovative efforts	OPAT	Number of ordinary patent applications filed in a given year in the state
4.	Infant mortality rate	IMR	Number of deaths of infants below one year old per 1000 live births in a given year for the state
5.	Gross enrolment ratio at upper primary level	GER_UPP	Ratio of total enrolment in grades VI–VIII to the population in the age group 11–13 years in the state (in percentage terms)
6.	Elementary education infrastructure	EEI	Percentage of schools with electricity connection in a given year in the state
7.	Schools with computers	SCH_CMP	Percentage of schools with computers in the state

Table 2 Descriptive statistics (number of observations = 54)

Sl.	Variables	Mean	Standard deviation	Sl.	Variables	Mean	Standard deviation
1	ILP	33.80	11.09	5	GER_UPP	85.60	11.11
2	IFCL	13.58	10.83	6	EEI	53.40	35.77
3	OPAT	426.29	617.73	7	SCH_CMP	26.66	23.67
4	IMR	39.24	12.30				

education infrastructure in terms of the proxy variable percentage of schools with electricity connection (EEI). This indicator may also capture the general availability of electricity in the state. Electricity is used for various economic activities in both rural and urban areas. Availability of electricity can increase production activity late in the evening and sometimes beyond mid night. Third variable is percentage of schools with computer (SCH_CMP). This measures the amount of importance given by the state to computer based education in schools, which may be quite important in today's skill biased development (Bhat and Siddharthan 2013).

Table 2 gives the descriptive statistics for the sample. The mean value of industrial labour productivity for the large states during the period of study (2011–2013) was around Rs. 34 Lakhs per worker. During all three years, Gujarat had the highest industrial labour productivity at more than 56 Lakhs per worker. For the large states, the average investment on fixed capital per worker in the industries was around Rs. 14 Lakhs. Of all the states in the study, Orissa had the highest fixed capital investment per worker during all the three years (2011–2013). Kerala and Bihar were among the states with lowest fixed capital investment per worker during the study period.

With regards to the health indicator, the average infant mortality rate for the large states was around 39. During all the three years, Madhya Pradesh had the highest value and Kerala had the lowest value on IMR. In the bigger states of India, the average gross enrolment ratio at upper primary level (GER_UPP) is around 86 %. During all three years, Tamil Nadu has the highest and Bihar has the lowest GER_UPP measure among the sample states. Only around 53 % of the schools on an average have electricity connection. Punjab and Gujarat have been the leading states in term of EEI indicator with more than 97 % of the schools having electricity connection. Bihar and Jharkhand are at the other extreme with less than 12 % of schools having electricity connection. The mean of percentage of schools with computers (SCH_CMP) is low at only 27 % with less than 2 % schools in Bihar and more than 85 % of the schools in Kerala having computers.

In the case of variable representing innovative efforts, OPAT, that average number of patents filed by the large states during the last three years is around 426. However, the standard deviation is also high. More detailed information on number of patents filed in each state is presented in Fig. 1. As is clear, Maharashtra is the

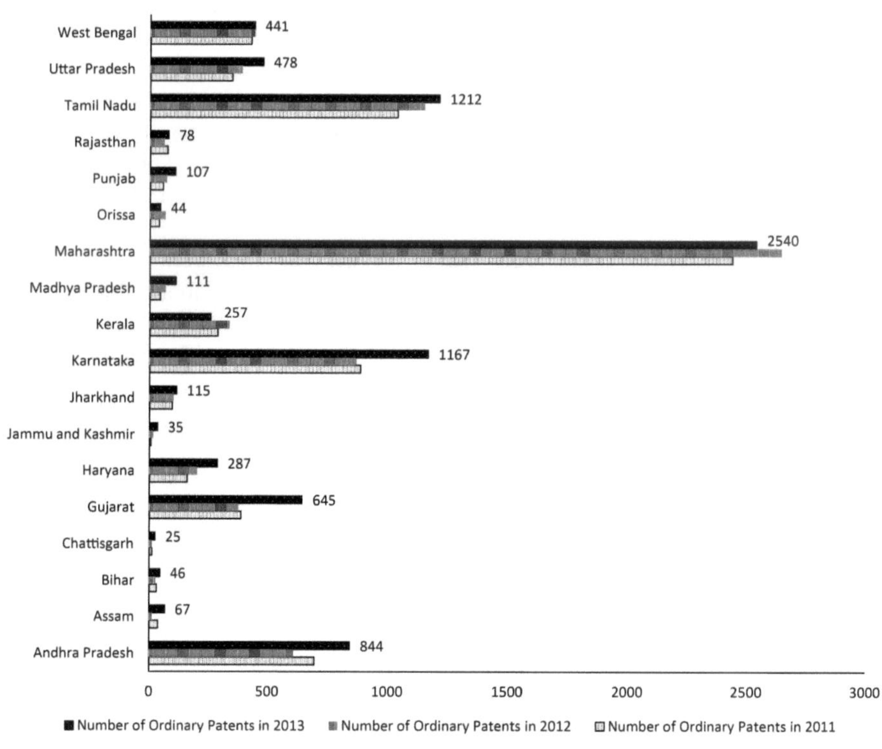

Fig. 1 Number of ordinary patents filed by the states during 2011–2013

leading state in terms of number of ordinary patents filed with around 2540 ordinary patents during 2013. During the same year, Tamil Nadu and Karnataka each account for 1212 and 1167 ordinary patents, respectively. Karnataka has many of the leading information technology companies like Samsung, Infosys and Wipro that are active in terms of filing patents. The states of Chattisgarh and Jammu and Kashmir have least number of patents filed in the state.

4 Econometric Models and Analysis

The data is a balanced panel data consisting of 18 states and 3 years. Hence, panel data regression model has been used for the analysis. Following Baltagi (2005), the econometric model is specified as follows:

$$y_{it} = \alpha + X_{it}'\beta + u_{it}$$

where y is the explained variable, namely, industrial labour productivity (ILP) and X is the vector of relevant explanatory variables from Table 1. The value of i ranges from 1, …, 18 (representing the states) and t takes the value 1–3 (for the years, 2011, 2012 and 2013 respectively). α is a scalar, β is coefficient on the explanatory variables and u is the stochastic error. Hausman specification test (Baltagi 2005) is used to choose between fixed and random effects model. Statistical analysis has been carried out in STATA version 10 statistical package. The coefficients of the explanatory variables representing innovative efforts (OPAT), capital (IFCL) and education related indicators (GER_UPP, EEI, SCH_CMP) are expected to have positive signs. The coefficient on the health indicator (IMR) is expected to have a negative sign.

Table 3 presents the correlation matrix for the variables used in this study. Industrial capital to labour ratio is correlated to industrial labour productivity. Variable representing innovative efforts (OPAT) is correlated to the human capital indicators. However, the correlation coefficient is less than 0.5. The health variable

Table 3 Correlation matrix

	ILP	IFCL	OPAT	IMR	GER_UPP	EEI	SCH_CMP
ILP	1.00						
IFCL	0.51*	1.00					
OPAT	0.19	−0.09	1.00				
IMR	0.15	0.35*	−0.49*	1.00			
GER_UPP	0.05	−0.04	0.32	−0.53*	1.00		
EEI	−0.02	−0.28	0.47*	−0.63*	0.51*	1.00	
SCH_CMP	−0.05	−0.30	0.35*	−0.77*	0.55*	0.83*	1.00

*Indicates statistical significance at 1 % level

Table 4 Determinants of inter-state differences in industrial labour productivity (ILP)

Regressors	Model 1	Model 2[a]	Model 3[a]	Model 4	Model 5	Model 6	Model 7
Constant	26.77 (9.33)*	31.79 (5.20)*	26.94 (7.12)*	24.90 (7.67)*	25.64 (7.91)*	31.56 (4.08)*	38.99 (6.18)*
IFCL	0.32 (3.64) *	0.32 (3.64)*	0.32 (3.59)*	0.31 (3.65)*	0.32 (3.65)*	0.30 (3.38)*	0.29 (3.21)*
OPAT	0.01 (2.16) **	0.01 (2.23)**	0.01 (2.01)**	0.005 (1.41)	–	0.005 (1.55)	–
IMR	–	–	–	–	–	−0.11 (−0.67)	−0.23 (−1.68) ***
GER_UPP	–	−0.06 (−0.93)	–	–	–	–	–
EEI	–	–	−0.004 (−0.07)	–	–	–	–
SCH_CMP	–	–	–	0.1 (1.35)	0.14 (2.15)**	–	–
No. of Obs.	54	54	54	54	54	54	54
Wald-Chi^2	18.30*	19.14*	17.97*	20.48*	18.14*	18.61*	15.88*

All the econometric models are random effects based on Hausman test. z-statistics is in parenthesis
[a]Even in the absence of OPAT, coefficients of GER_UPP and EEI are statistically insignificant in determining ILP
*, **, ***Represent statistical significance at 1, 5 and 10 % respectively

IMR is correlated to industrial fixed capital-labour ratio and has a high correlation coefficient of above 0.5 with all the three education related variables.

Table 4 presents the results for the econometric models with various combinations of the explanatory variables. As expected, fixed capital to labour ratio is statistically significant in determining industrial output in all the seven econometric models. Thus, higher amounts of investments on fixed capital like modern plants and machinery can improve industrial worker's output.

The variable representing innovative efforts (OPAT) is statistically significant with a positive sign in Model 1. Thus, higher innovative efforts by the firms and researchers in the state do imply better labour productivity of industrial workers in the states. In this study, the large states with higher values on patents happen to be Maharashtra, Tamil Nadu and Karnataka. In the light of the innovation systems approach, there is evidence that states like Maharashtra do have institutions that support small and medium enterprises (SMEs) in terms of funding, helping in filing patents and even building intellectual property (FICCI 2012). Often there can be spill-over effects of innovative activities in the region.

Even in the presence of education based variables like GER_UPP (Model 2) and EEI (Model 3), the variable representing innovative efforts is statistically significant in determining labour productivity of the industries. However, in the presence of

SCH_CMP (Model 4) and IMR (Model 6), the variable loses its statistical significance. This may be due to the presence of correlation between OPAT and the other two variables.

In the absence of OPAT, of the three indicators related to education, only SCH_CMP is important in determining inter-state differences in industrial output (Model 5). When the school children are exposed to latest technologies like computer, their skills improve. The households of these children also get exposed to computer and other latest technologies including information and communication technologies. This in turn may help them bring in efficiencies in their work and also make more informed decisions with regards to economic activities. Furthermore, familiarity with internet makes children motivated for higher education and better employment.

Again, in the absence of OPAT, the health indicator IMR is statistically significant with expected sign (Model 7). A healthy population in general implies more productive labour and hence higher industrial output in the state.

5 Conclusion and Implications

This study tried to understand the factors responsible for inter-state differences in industrial labour productivity in India. The study specifically explored the role of innovative efforts, in terms of patent applications filed, in determining the differences in industrial output per worker for the bigger states in India. In line with the theory on production function, this study empirically confirmed that choice of technology in terms of capital-labour ratio is important in determining industrial labour productivity. After accounting for capital, innovative efforts are also important in determining inter-state differences in industrial labour productivity. Thus, one can further explore the role of strong institutions and networks in promoting innovative efforts by organizations in the state, which ultimately can lead to higher industrial output per worker in the state.

Human capital variables like IMR and percentage of schools with computers also determine industrial output per worker. Healthy and skilled workforce can contribute towards improved labour productivity (Bhat and Siddharthan 2013. In line with other studies carried out in context of other countries (Tengtrakul and Peha 2013), in India too, one can further explore the spill-over effects of introduction of computers in schools on the economic activities and decisions of the households, which ultimately will affect the economic development of the state.

Appendix I

Sources of data

Sl.	Measure	Source
1.	State-wise value of industrial output in Rs. Crore	Annual Survey of Industries
2.	State-wise fixed capital in Rs. Crore	Annual Survey of Industries
3.	State-wise number of workers	Annual Survey of Industries
4.	State-wise ordinary patent applications filed	Annual reports of The Office of the Controller General of Patents, Designs, Trademarks and Geographical Indications, Department of Industrial Policy and Promotion, Ministry of Commerce and Industry, Government of India
5.	State-wise infant mortality rate	Sample Registration System (SRS) Bulletins, Office of the Registrar General and Census Commissioner, Ministry of Home Affairs, Government of India
6.	State-wise gross enrolment ratio at upper primary level	District Information System for Education (DISE), National University of Educational Planning and Administration (NUEPA), New Delhi, India
7.	State-wise percentage of schools with electricity connection	District Information System for Education (DISE), National University of Educational Planning and Administration (NUEPA), New Delhi, India
8.	State-wise percentage of schools with computers	District Information System for Education (DISE), National University of Educational Planning and Administration (NUEPA), New Delhi, India
9.	Whole sale price index with 2004–05 = 100 as base	Office of Economic Adviser, Department of Industrial Policy and Promotion (DIPP), Ministry of Commerce and Industry, Government of India

References

Agarwalla, A., & Pangotra, P. (2011). Regional income disparities in India and test for convergence—1980 to 2006, W.P. No. 2011-01-04, working paper series, Indian Institute of Management Ahmadabad, pp. 1–22.

Akçomak, I. S., & ter Weel, B. (2009). Social capital, innovation and growth: Evidence from Europe. *European Economic Review, 55*, 544–567.

Baltagi, B. H. (2005). *Econometric analysis of panel data* (3rd ed.). England: Wiley.

Banerjee, R., & Roy, S. S. (2014). Human capital, technological progress and trade: What explains India's long run growth? *Journal of Asian Economics, 30*, 15–31.

Barro, R. J. (1991). Economic growth in a cross section of countries. *The Quarterly Journal of Economics, 106*(2), 407–443.

Barro, R. J. (2001). Human capital and growth. *American Economic Review, 91*(2), 12–17.

Bhat, S., & Siddharthan, N. S. (2013). Human capital, labour productivity and employment. In N. S. Siddharthan & K. Narayanan (Eds.), *Human capital and development: The Indian experience* (pp. 11–22). New Delhi: Springer.

Chandrasekhar, S., & Sharma, A. (2014). Internal migration for education and employment among youth in India, WP-2014-004, working paper, Indira Gandhi Institute of Development Research.

Cooke, P., Uranga, M. G., & Etxebarria, G. (1997). Regional innovation systems: Institutional and organisational dimensions. *Research Policy, 26*, 475–491.

FICCI. (2012). Innovation readiness of Indian SMEs: Issues and challenges, FICCI MSME Summit 2012 knowledge paper, Federation of Indian Chambers of Commerce and Industry.

Freeman, C. (1995). The 'National System of Innovation' in historical perspective. *Cambridge Journal of Economics, 19*, 5–24.

Gregorio, J. D., & Lee, J. (2002). Education and income inequality: New evidence from cross-country data. *Review of Income and Wealth, 48*(3), 395–416.

Grossman, G. M., & Helpman, E. (1994). Endogenous innovation in the theory of growth. *Journal of Economic Perspectives, 8*(1), 23–44.

Howitt, P. (2000). Endogenous growth and cross-country income differences. *The American Economic Review, 90*(4), 829–846.

Knowles, S., & Owen, P. D. (1995). Health capital and cross-country variation in income per capita in the Mankiw-Romer-Weil model. *Economic Letters, 48*, 99–106.

Lin, S. (2015). Are ivory towers truly ivory? Knowledge spillovers and firm innovation. *Journal of Economics and Business, 80*, 21–36.

Lundvall, B.-Å. (Ed.). (1992). *National innovation systems: Towards a theory of innovation and interactive learning*. London: Pinter Publishers.

Maiti, D., & Mitra, A. (2010). Skills, informality, and development, IEG working paper no. 306.

Maitra, B., & Mukhopadhyay, C. K. (2012). Public spending on education, health care and economic growth in selected countries of Asia and the Pacific. *Asia-Pacific Development Journal, 19*(2), 19–48.

Mo, D., Swinnen, J., Zhang, L., et al. (2013). Can one-to-one computing narrow the digital divide and the education gap in China? The case of Beijing migrant schools. *World Development, 46*, 14–29.

Nelson, R. R. (Ed.). (1993). *National innovation systems: A comparative analysis*. Oxford: Oxford University Press.

Nelson, R. R., & Winter, S. G. (1982). *An evolutionary theory of economic change*. Cambridge, Massachusetts: The Belknap Press of Harvard University Press.

Ogburn, W. F., & Allen, F. R. (1959). Technological development and per capita income. *The American Journal of Sociology, 65*(2), 127–131.

Oketch, M. O. (2006). Determinants of human capital formation and economic growth of African countries. *Economics of Education Review, 25*, 554–564.

Oosterbaan, M. S., de Ruyter van Steveninck, T., & van der Windt, N. (Eds.). (2000). *The determinants of economic growth*. New York: Springer Science + Business Media.

Paweenawat, S. W., & McNown, R. (2014). The determinants of income inequality in Thailand: A synthetic cohort analysis. *Journal of Asian Economics, 31–32*, 10–21.

Purohit, B. C. (2008). Health and human development at sub-state level in India. *The Journal of Socio-Economics, 37*, 2248–2260.

Rutten, R., & Boekema, F. (2007). Regional social capital: Embeddedness, innovation networks and regional economic development. *Technological Forecasting and Social Change, 74*, 1834–1846.

Solow, R. M. (1956). A contribution to the theory of economic growth. *Quarterly Journal of Economics, 70*(1), 65–94.

Tengtrakul, P., & Peha, J. M. (2013). Does ICT in schools affect residential adoption and adult utilization outside schools? *Telecommunications Policy, 37*, 540–562.

Verspagen, B. (2005). Innovation and economic growth. In J. Fagerberg, D. C. Mowery, & R. R. Nelson (Eds.), *The Oxford handbook of innovation*. Oxford: Oxford University Press.

Patenting and Technical Efficiency of Manufacturing Firm in India: Evidence from Medium and High Technology Firms

Sunil Kumar and Ruchi Sharma

1 Introduction

Innovation is regarded as an output, resulting from inputs, where physical capital, human capital, R&D, and economies of scale all play major roles (Mokyr 2010). Recently, due to increase in availability of firm level data as well as more advanced methods, there is an increase in studies that estimate link between innovation and performance of firms (Bartelsman and Doms 2000). Existing literature used two measures of innovative activity, R&D spending and patent counts (Hall 2011). Some researchers explore the influence of investment in knowledge capital like R&D on the performance of firms (Mansfield 1984; Griliches and Mairesse 1984; Raut 1995). Some other researchers argue that innovation in the form of patent contributes to the higher performance of firms (Deolalikar and Roller 1989; Eaton and Kortum 1996; Crepon et al. 1998). The present study, however, considers the relationship between innovation and technical efficiency (TE) of Indian manufacturing firms.

The study estimate TE through Data Envelopment Analysis (DEA) which is a frontier non parametric estimation method. TE indicates the capability of a firm to attain maximum output from a given set of input. Since patent data includes product and process innovation, efficiency improvements and variation among the firms can

This chapter is an extension of the authors' work "Impact of patenting on firms' performance: an empirical investigation based on manufacturing firms in India" that was published in 2016 in the *Economics of Innovation and New Technology*, Vol. 25, Issue 1.

S. Kumar (✉)
School of Business Studies and Social Sciences, Christ University,
Bennarghetta Road Campus, Bangalore 560076, India
e-mail: sunilkumar.a@christuniversity.in; sunileco30@gmail.com

R. Sharma
School of Humanities and Social Sciences, Indian Institute of Technology, Indore, India
e-mail: ruchi@iiti.ac.in; ruchi.shm@gmail.com

N.S. Siddharthan and K. Narayanan (eds.), *Technology*,
India Studies in Business and Economics,
DOI 10.1007/978-981-10-1684-4_3

be attributed to the innovative activities of firms. More likely foreign firms are more superior in firm specific assets and managerial capability. Therefore, a distinction between foreign and domestic firms helps to distinguish the efficiency discrepancy among them. We adapt the idea of group frontier from Bhandari and Ray (2012) who construct different production possibility set for different groups based on location, ownership and organizational forms. The present study however constructs the technological frontier based on their technology content.[1] In order to cross check the results the study also estimates technical efficiency based on meta-frontier approach, which construct a common technology frontier for all firms. In this paper, we use a panel of 489 firms for high and medium technology sectors during 2000–2010 to estimate the technical efficiency of firms due to patenting. The study finds that patenting by firms has an influence on their efficiency levels whereas the role of R&D is insignificant. Further, majority of Indian firms are suboptimal but there is an indication of overall efficiency improvement of the firms in the sector.

The rest of the study is organized into 5 sections. Section 2 provides some theoretical linkages and empirical evidence of the existing studies. Section 3 provides an overview of the research methodology applied in the study. Sections 3.1 and 3.2 provide the brief introduction of DEA and Tobit analysis respectively. Section 4 focuses on the description of data and variables used in the study. Section 5 presents the results and analysis. Section 6 offers some concluding remarks.

2 The Theoretical Linkage and Empirical Evidence

Technological progress evolves as a step by step process. In the first attempt, lagging firms in an emerging economy needs to catch-up with the existing technology frontier before race for the technological leadership in the future (Aghion et al. 2001). The present study therefore makes an attempt to evaluate the role of innovation in closing the gap of a lagging firm with the technological frontier in the industry. The study assumes that the rate at which technological gap closes is associated with the ability of a lagging firm to receive the technology flows from frontier. In this particular case, a firm's investment function acts as an engine to bridge the gap between frontier and the follower. Assessing the technical efficiency of a firm is the prevailing method for evaluating the technological gap and performance of a firm. Technical efficiency is the ratio of the actual output to the maximum producible output and it is attained when output is optimized from a given set of inputs. Theories on technological innovation at the firm level consider

[1]OECD defines manufacturing firms into high tech, medium tech, medium low tech and low technology sectors based on their technology content. The present study includes only high and medium technology sectors and the list is given in Appendix A.

both research input in the form of R&D investment and output in the form of patent to contribute towards the higher performance (Kamien and Schwartz 1982; Griliches 1987; Crepon et al. 1998). There are several studies that examine the link between innovation and productivity of firms (Griliches and Mairesse 1984; Deolalikar and Roller 1989; Coe and Helpman 1993; Crepon et al. 1998; Griffith et al. 2006). However, there are a few studies that examine the link between innovation and technical efficiency. Bremmer et al. (2008) have analyzed the impact of innovation on the technical efficiency of Dutch firms during 1991–1998, but the study failed to isolate any specific impact of innovation on the technical efficiency. By using a recent data set in Spanish manufacturing firms Sanchez and Diaz (2013) have shown that large innovative firms are more efficient than non innovators. In a more recent attempt, Mayans and Perez (2014) included the process and product innovation variables in the estimation of production frontier and find that both innovations increase the efficiency of firms. The present study therefore, empirically examines whether the innovative activities of medium and high tech firms in India have any influence on their efficiency improvement.

3 Research Methodology

Akin to Luoma et al. (1996) and Fethi et al. (2000) the present study follows a two-stage approach, in which the first stage involves the specification and estimation of technical efficiency and in the second stage we use Tobit regression model to explain the technical efficiencies. In the first stage, we rely on DEA based Malmquist Productivity Index (MPI) to estimate technical efficiency. DEA is a frontier non-parametric analytical tool that empirically measures the relative efficiency of a decision making units (DMUs).[2] The frontier approach assumes that, there exists a best practice production function corresponding to the set of maximum attainable output levels for a given combination of inputs. Meeusen and Van den Broeck (1997) have independently contributed to the econometric modeling of the frontier production function and paved the way for theoretical research and empirical application of the model to many different industries. Stochastic frontier approach is the other way to estimate the productivity under frontier analysis.[3]

We select DEA as an appropriate method for the present study due to the following reason. First, it does not require any functional specification as it looks for a best practice frontier within the data rather than considering the average path

[2]Farell (1957) originally developed the idea of measuring efficiency as a relative distance from frontier through a non parametric approach. Later, Charnes et al. (1978) extended the frame work and gave the name DEA.

[3]In the frame work of frontier and non-frontier approaches, researchers and policy makers employed parametric, semi-parametric and non-parametric method to analyze the productivity and efficiency of different Decision Making Units (DMUs). Mahadevan (2003) and Kathuria et al. (2013) provides a detailed explanation of these approaches.

through the middle points of a series of data. Second, DEA is appropriate when the relative importance of various inputs and output cannot be defined. In our case, absence of available market prices limits allotting the respective weights to various inputs in the production process. DEA however, overcome this problem by allowing each DMU to choose the vectors of input and output weights, which maximizes its own ratio of weighted output to weighted inputs subject to the constraints. Finally, the decision to choose frontier or the non frontier approach depends on the research question. If the intention of a researcher is to estimate the contribution made by each factor of production to productivity, the non frontier approach is considered as superior to frontier analysis. Contrary to this, the present study estimates the best practice output level only and, therefore, frontier analysis is considered as the more appropriate method.

The above explained special features of DEA have acquired attraction from many researchers (Korhonen et al. 2001; Wang and Huang 2007). In Indian context, the DEA method is popularized after Ray (2002), who used the method to estimate the technical efficiency of firms after the reform period. Since then, there are numerous studies that employ DEA technique to estimate the efficiency in Indian manufacturing (Kumar 2005; Marjit and Kar 2009; Raj 2011; Raj and Babu 2011).

3.1 An Illustration of Data Envelopment Analysis

Firms normally seek to increase their market share which enables them to acquire some degree of monopoly power. Therefore, the present study considers output maximization as a firm's objective. We consider the conventional DEA framework where labour and capital are the two inputs and deflated sale is the output for the study. DEA constructs an efficient frontier composed of those firms that produce as much output as possible from given input levels. Firms at the frontier are efficient while those below are inefficient. Further, the efficiency score depends on how well a DMU performs over other firms. A brief illustration of the construction of DEA optimization is given in Appendix B.

3.2 Tobit Estimation

In the second stage, the existing studies follow censored regression model (Tobit model) since the dependent variable is limited in between 0 and 1 (Luoma et al. 1996; Gillen and Lall 1997). In this stage, these estimated efficiency scores are explained through a set of variables those are expected to influence the performance of firms. Following this approach, the present study also applied Tobit estimation for explaining the efficiency of Indian high and medium-high tech firms. Tobit model is a censored normal regression model, censoring from below at zero (Cameron and Trivedi 2010). The underlying model is:

$$y^* = X^! \beta + \varepsilon \tag{1}$$

where y* is a continuous latent variable ranging between 0 and 1, X is a matrix of explanatory variables, β is a vector of the coefficients to be estimated and ε is a vector of error terms which is normally distributed with mean 0 and variance σ^2. Since our dependent variable is censored from both below and above we apply a Tobit model (Cameron and Trivedi 2010). Hence, our model will become:

$$\begin{aligned} y &= 0 \quad \textit{if}\, y^* \le 0 \\ y &= y^* \quad \textit{if}\, 0 < y^* < 1 \\ y &= 1 \quad \textit{if}\, y^* \ge 1 \end{aligned} \tag{2}$$

4 Data and Variable Description

The research setting of the study is the firms from medium and high-technology industries in India, the relative research intensive sectors. Particularly, we focus on firms which are producing consistent data on input and output variables. The main source of data for our study is the website of Controller General of Design, Trademark and Patent (CGDTP). All patent data used in this study are based on patent granted to each firm at the Indian Patent office (IPO). We arrange all the patents based on the application date on the assumption that there is no time lag between patent application and completed invention. Center for Monitoring Indian Economy (CMIE) prowess data base is used to collect all firm level information and all firms' variables are deflated with appropriate deflators. Sales data is deflated by industry specific Whole Sale Price (WPI) index. The index is obtained from the website of Office of Economic Advisor (OEA), the Ministry of Commerce and Industry. The capital data is deflated by the capital deflator and index number of industrial production is obtained from Handbook of Statistics on Indian Economy (RBI).

4.1 Input and Output Variables

In this study, similar to Mahadevan (2001) one output and two input variables are used. Deflated net sale, sales net of indirect taxes, is the output variable for DEA. The argument in favor of net sale is that indirect taxes are imposed by government and therefore not reflected in the productive capacity or the operation of the company. We consider two conventional inputs namely, labour and capital for the DEA. For capital we consider Gross Fixed Asset (GFA) that comprise of net of tangible assets, land and buildings, plant and machinery, transport and communications, furniture and other fixed assets. The advantage of using GFA as a capital is that the value is given in stock format and allowed for normal depreciation. Since we do not have direct information of labours, we constructed the labours information following (Sharma 2010). We obtain average wage rate (wages for the workers/number of workers) of each industry from the Annual Survey of Industries (ASI) database and each firm's salaries and wages divided by the average wage rate obtained from ASI, which provides the number of labour employed by firm.[4]

4.2 *Determinants of Efficiency*

The technical efficiency scores obtained from the first stage regression are explained through relevant variables that are likely to affect the efficiency levels across the firms. Based on the extensive literature survey, we identified the variables that can explain the technical efficiency of manufacturing firms. Similar to Deolalikar and Roller (1989) and Crepon et al. (1998) we consider patenting by firms as the major determinant of firms differences in efficiency. The study considers stock of patent as a measure of countries knowledge stock. In our study, we adapt the concept from Chang et al. (2013) to construct the firm level knowledge stock (PAT) through their grated patent. Firms invest in R&D activities by expecting the product, process or organizational innovation, all of which together contribute to the higher performance of firms. Hence, we consider R&D stock (RDS) as one of the explanatory variables that determines the level of efficiency of firms.[5, 6] Not only firm's own R&D but technology purchased from others also enhances productivity of firms (Basant and Fikkert 1996). Therefore, we introduce licensing payment (LIC) made by a firm as an additional explanatory variable.

Many researchers consider the impact of foreign ownership (FOS) on the level of efficiency and performance of firms (Aitkin and Harrison 1999; Sharma 2010). This is because foreign firms have superiority in firm specific asset and management capability which are necessary to bring them to the frontier technology. These studies however do not claim that foreign ownership is an indisputable element in the catching up process.

---

[4]ASI provides average wage rate (AWR) (total emoluments/total man days) for each 3 digit industries. We have information of each firm's salaries and wages. By assuming that in each industry, the AWR would be same, we divide each firm's salaries and wages with respect to AWR. This provides the information of total number of labors employed.

[5]R&D stock has been constructed through perpetual inventory method, the process outlined by Levinsohn and Petrin (2003). Specifically, we calculate R&D stock as $R_t = I_t + (1 - \delta)R_{t-1}$ where R_t = stock of R&D, I_t R&D investment made at time t and δ is the 15 % depreciation rate..

[6]In order to check the likely occurrence of multicollinearity in between R&D and patents, we introduce these variables separately in the model.

5 Results and Analysis

5.1 DEA Results

Table 1 reports the frequency distribution of TE calculated annually for each sector. The first column represents number of firms in each sector along with the group name. Considering TE, we categories the firm as optimal, suboptimal and least efficient depending upon the efficiency score equal to 1, between 0.99 and 0.75 and less that 0.74 respectively. The analysis shows that most firm for all sectors are in the range of 0.99–0.75, reflecting that inefficient firm may not sustain for long in the industry. This result should be understood in view of the data cleaning process that we followed where we removed all loss making units. The medical and motor sectors are the most efficient with 62.5 and 60 % of firms respectively in the suboptimal category in 2000. However, we will not use data on motor and medical sectors for further analysis as the number of firms in this case is small. In case the minimum number of firms chosen for DEA is less than 3 times of total number of inputs and outputs, the probability of an inefficient firm being declare as efficient increases. Therefore, it is not desirable to compare the efficiency levels of firms across industries as the number of firms in each sector vary between 5 and 159. However, a comparison of industries over the years is possible as the number of firms remain constant. The analysis shows efficiency improvement in chemical sector where the number of efficient firms doubles in 2010 compared to 2000. Similarly, firms in pharma sector have also improved their efficiency during the same period. One positive sign in all the sectors is that the firms with the efficiency level below 0.75 has been reduced significantly during the study period. The study further investigates the efficiency differential of the foreign and domestic firms and the results are given in Table 4.

Table 2 reports annual average TE of various sectors during 2000–2010 estimated through group frontier. The overview shows that Indian medium and high-tech firms are technically efficient at the 87 % level (last column of the Table 2). Their overall efficiency level of 87 % implies that, if they would have been technically efficient, they could have produced 13 % of more output from the given level of inputs. Among the sectors R.T.C. is the highest efficient as they realized 95 % of their potential output from the given inputs (in the year 2005 and 2006), while pharma is the least efficient with 83 and 84 % realization of their potential output during the same years. One probable reason to the highest efficiency of RTC is the small number of firms in that group. Since DEA is a relative measurement of efficiency, variation in number of firms in each category may lead to variation in the calculation of efficiency. However, except in electrical sector, there is an efficiency increment in all the sectors over the years. On an average, technical efficiency of manufacturing firms in India has increased by 3 %. Specifically, annual average technical efficiency of 0.85 in 2000 has increased in 0.88 in 2010.

Table 3 gives annual average TE of various sectors during 2000–2010 estimated through meta-frontier. As against to the previous method, meta-frontier result

Table 1 Frequency distribution of technical efficiency scores

Sectors	Range of firms	Year										
		2000	01	02	03	04	05	06	07	08	09	2010
Pharma	1	7	11	11	9	10	8	8	9	7	7	11
(83)	0.75–0.99	56	55	62	59	57	57	58	57	58	60	62
	<0.74	20	17	10	15	16	18	17	17	18	16	10
RTC	1	8	9	9	7	7	10	12	11	7	7	7
(19)	0.75–0.99	10	8	8	9	10	8	7	7	10	11	11
	<0.74	1	2	2	3	2	1	0	1	2	1	1
Medical	1	5	5	5	5	4	4	4	6	4	3	3
(8)	0.75–0.99	3	3	3	3	4	4	4	6	4	3	3
Electrical	1	11	11	8	8	7	7	8	11	10	9	10
(42)	0.75–0.99	27	28	32	31	31	30	30	26	30	33	28
	<0.74	4	3	2	3	4	5	4	5	2	0	4
Motor	1	3	3	3	3	3	3	3	3	3	3	3
(5)	0.75–0.99	2	2	2	2	2	2	2	2	2	2	2
Chemical	1	6	8	9	12	11	14	13	14	12	12	12
(159)	0.75–0.99	117	122	117	126	129	133	132	132	135	137	135
	<0.74	36	29	33	21	19	12	14	13	12	10	12
Rail	1	6	6	10	11	8	8	7	4	7	6	7
(86)	0.75–0.99	76	76	76	75	77	78	78	81	77	78	78
	<0.74	4	4	0	0	1	0	1	1	2	2	1
Machinery	1	7	7	9	10	10	7	8	8	8	10	7
(87)	0.75–0.99	55	55	56	60	68	68	65	61	64	67	71
	<0.74	25	25	22	17	9	12	14	18	15	10	9

Note Number in parenthesis represent total number of firms in each sector

Table 2 Annual and sectoral average of technical efficiency (through group frontier)

Year	Chemical	Electrical	Machinery	Pharma	Railroad	RTC	ATE
2000	0.82	0.9	0.81	0.82	0.85	0.91	0.85
2001	0.83	0.89	0.81	0.84	0.85	0.9	0.85
2002	0.82	0.89	0.82	0.86	0.9	0.91	0.87
2003	0.85	0.88	0.84	0.84	0.9	0.9	0.87
2004	0.85	0.89	0.85	0.85	0.89	0.91	0.87
2005	0.86	0.88	0.84	0.83	0.89	0.95	0.87
2006	0.87	0.9	0.84	0.84	0.9	0.95	0.88
2007	0.87	0.91	0.84	0.84	0.9	0.94	0.88
2008	0.85	0.9	0.85	0.85	0.89	0.91	0.87
2009	0.87	0.89	0.85	0.85	0.89	0.9	0.88
2010	0.87	0.89	0.86	0.86	0.89	0.92	0.88
Mean	0.85	0.89	0.84	0.84	0.88	0.92	0.87

Table 3 Annual and sectoral average of technical efficiency (through mega frontier)

Year	Chemical	Electrical	Machinery	Pharma	Rail	RTC	ATE
2000	0.80	0.79	0.80	0.81	0.76	0.89	0.82
2001	0.71	0.76	0.79	0.76	0.65	0.78	0.73
2002	0.74	0.80	0.78	0.79	0.78	0.67	0.73
2003	0.81	0.92	0.76	0.83	0.74	0.93	0.85
2004	0.79	0.74	0.79	0.75	0.80	0.72	0.78
2005	0.80	0.82	0.78	0.73	0.77	0.85	0.78
2006	0.77	0.80	0.80	0.78	0.78	0.83	0.79
2007	0.85	0.90	0.79	0.79	0.87	0.89	0.84
2008	0.87	0.91	0.89	0.88	0.88	0.84	0.89
2009	0.77	0.78	0.83	0.76	0.78	0.88	0.78
2010	0.81	0.87	0.84	0.81	0.83	0.86	0.84
Mean	0.79	0.83	0.80	0.79	0.79	0.83	0.80

Table 4 Technical efficiency: foreign versus domestic firms

	Foreign		Domestic		Foreign		Domestic	
Year	2000	2010	2000	2010	2000	2010	2000	2010
Chemical	0.81	0.85	0.79	0.80	0.85	0.90	0.81	0.86
Electrical	0.75	0.92	0.80	0.87	0.93	0.92	0.89	0.89
Machinery	0.80	0.84	0.80	0.84	0.79	0.86	0.82	0.86
Pharma	0.80	0.77	0.81	0.81	0.89	0.93	0.81	0.85
RTC	0.90	0.94	0.88	0.86	0.91	1.00	0.91	0.91
Rail	0.78	0.82	0.75	0.84	0.88	0.91	0.84	0.89
Method	Mega	Mega	Mega	Mega	Group	Group	Group	Group

shows that the overall efficiency level is 80 %. The possible differences are expected because firms from different sectors face different technologies and constructing a single frontier based on all firms may be an inappropriate technology frontier. The table shows that in the present case, RTC is the highly efficient sector. Prima facie, we do not find any significant difference between group frontier and meta- frontier while estimating TE.

Table 4 gives comparative performance of foreign and domestic firms in terms of their efficiency scores. While analyzing the individual sectors both methods of estimation shows that foreign firms outperform their domestic counterparts.

5.2 *Tobit Estimation Result*

As we are dealing with panel data, the issues like heteroskedasticity and endogenity emerge and we have addressed these issues in our modeling. Appendix C shows that likelihood ratio tests confirm the problem of heteroskedasticity and the bootstrap standard errors minimize the problem. The Hausman Taylor endogenity test does not confirm the endogenity issue and the results of the tests are given in Appendix D. In order to test the potential colinearity among the explanatory variables we perform a correlation test and the result is given in Appendix E and it shows that there is no high correlation among the explanatory variables.

For best approximation in the result, we use the fixed effect censoring Tobit estimation developed by Honore (1992).

Results of panel Tobit estimation are suggested by Honore (1992) are given in Table 5. Estimation through group frontier is given through columns I–III and estimation through meta-frontier is given through columns IV–VI. We employ three different equations to estimate the influence of R&D and patenting on the efficiency of firms. Column I &IV use both RDS and PAT variables simultaneously whereas rest of the columns uses these variables separately to check for multicollinearity. The results indicate a significant impact of patenting on the efficiency differential among firms. This shows that firms having large patent pool show higher efficiency than others. The result however does not produce any significant impact of R&D on technical efficiency. One probable reason is that there may be a considerable time lag between R&D spending and the happening of the results. Therefore, the benefit of R&D will observe only at a later time period. We have repeated the same experiment through fixed effect and random effect model. The results of the same has been given in Appendix F. Fixed effect and random effect models also produce significant impact of patenting on technical efficiency, whereas fails to produce any impact of R&D. The result further indicates that technology purchased from outside has a positive and significant impact on the efficiency differential among firms. In the preliminary examination, we find that foreign firms are outperforming the domestic firms. The result however does not hold in the econometric examination. Here we find that there is no statistical difference between foreign and domestic firms in terms of technical efficiency.

Table 5 Determinants of technical efficiency

	Estimation through group frontier			Estimation through meta frontier		
Group	I	II	III	IV	V	VI
PAT	0.018 (0.009)	NA	0.019 (0.009)	0.053 (0.024)	NA	0.053 (0.026)
	[2.02]**		[2.01]**	[2.16]**		[2.11]*
RDS	NA	−0.002 (0.003)	−0.002 (0.004)	NA	−0.002 (0.008)	−0.003 (0.008)
		[−0.7]	[−0.55]		[−0.29]	[−0.36]
LIC	0.179 (0.013)	0.179 (0.016)	0.179 (0.015)	0.129 (0.031)	0.129 (0.028)	0.129 (0.028)
	[13.93]*	[11.47]*	[11.47]*	[4.12]*	[4.57]*	[4.56]*
FOS	0.002 (0.009)	0.002 (0.007)	0.002 (0.006)	−0.002 (0.012)	−0.003 (0.01)	−0.003 (0.01)
	[0.26]	[0.3]	[0.34]	[−0.25]	[−0.3]	[−0.28]
Year trend	0.002 (0.001)	0.002 (0.001)	0.002 (0.001)	0.009 (0.001)	0.009 (.001)	0.009 (0.001)
	[4.55]*	[7]*	[4.5]*	[16.98]*	[16.19]*	[16.15]*

Note NA indicates Not Applicable. Standard errors are given in parenthesis. 'z' scores are in square bracket
*, **Represent 1 and 5 % level of significance

6 Conclusion

The objective of the study is to examine the differences in efficiency due to innovative activities of firms. The study discusses about efficiency improvement in firms as a function of its investment in innovation. Investment in R&D expenditure and patent counts are considered as the indicators of innovation. The present study follows a two-stage approach, in which the first stage involves the specification and estimation of technical efficiency and in the second stage we use Tobit regression model to explain the causal determinants of technical efficiencies. Estimated technical efficiency scores based of DEA show an efficiency improvement during the study period. Investment in R&D and stock of patent respond differently to the efficiency improvements. Firms having large patent pools show higher efficiency whereas in earlier case it does not have any impact on the performance. The distinction between in house R&D and purchased technology is analyzed in the study and the technology purchase has found to have a significant impact on efficiency. Therefore, it requires a comprehensive study based on the role of different forms of technology on efficiency improvement. The study concentrated on efficiency aspect only and does not consider the issue of productivity improvement of firms due to patenting. This part is left for future study.

Appendix A: List of Sectors According to Their National Industrial Classification (NIC)

	Sectors	NIC 3 and 4 digit
High tech	Aircraft and spacecraft	3315
	Pharmaceuticals	210
	Office, accounting and computing machinery	262, 332, 2817
	Radio, TV and communications equipment	261, 63, 264, 322, 331, 3314
	Medical, precision and optical instruments	325, 266, 267, 332, 2651
	Electrical machinery and apparatus, n.e.c.	271, 272, 273, 274, 279, 3312, 3314, 332
Medium high-tech	Motor vehicles, trailers and semi-trailers	291, 292, 293, 3313
	Chemicals excluding pharmaceuticals	201, 202, 203, 268, 1079
	Railroad equipment and transport equipment	302, 309, 3315
	Machinery and equipment, n.e.c.	281, 282, 275, 252, 304, 3311, 3312, 3320, 2593.excl 2817

Appendix B: DEA Optimization Problem: An Algebraic Formulation for 1 Output and N Number of Inputs

Let y^k be the output and x_i^k (where i = 1, 2, …, n) be the inputs of the firm k (where k = 1, 2, …, m). We wish to measure the technical efficiency of firm E. The linear programming problem for the DEA would be;

$$\begin{aligned} &\max(\varphi), \textit{the objective function} \\ s.t.\quad &\sum_{k=1}^{m} \lambda_k x_{ik} \leq x_{iE} (i = 1, 2, \ldots n) \\ &\sum_{k=1}^{m} \lambda_k y_k \geq \varphi y_E \\ &\sum_{k=1}^{m} \lambda_k = 1; \lambda_k \geq 0 (k = 1, 2, \ldots m); \varphi \end{aligned} \tag{d1}$$

The technical efficiency of firm E would be

$$\tau_E = \frac{1}{\varphi^*} \qquad \text{(d2)}$$

where φ^* is the optimal solution for the linear programming problem.

Appendix C: Test for Heteroskedasticity

	Hetero panel	Non hetero panel
PAT	−0.003 (−0.4)	0.021 (1.01)
RDS	−0.003 (−0.27)	0.006 (0.84)
LIC	0.152 (10.22)	0.221 (8.14)
HHI	0.19 (13.49)	0.28 (9.51)
FOS	−0.002 (−0.97)	0.007 (1.81)
Constant	0.86 (736.27)	0.845 (367.57)
Likelihood ratio	$\text{Chi}^2(1) = 3404.83^*$	

Z scores are in parenthesis
'*' indicates 1 % level of significance

Appendix D: Hausman Taylor Test for Endogenity

	Htaylor	FEM
TV exogenous		
LIC	0.001 (1.49)	0.001 (0.6)
HHI	−0.021 (−0.7)	−0.061 (−1.97)
FOS	0.002 (0.39)	0.001 (0.13)
Year trend	0.003 (16.18)	0.004 (16.44)
TV endogenous		
PAT	−0.0001 (−0.89)	−0.0001 (−1)
RDS	−0.0003 (−2.15)	−0.0003 (−2.37)
TI exogenous		
ID	0.004 (0.38)	
Constant	−6.1123 (−14.19)	−6.232 (−14.44)

Note TV indicates varying and TI indicates time in varying variables. FEM means Fixed Effect Model

Appendix E: Correlation Matrix

	PAT	RDS	LIC	FOS
PAT	1			
RDS	0.418	1		
LIC	−0.039	−0.004	1	
FOS	0.060	0.002	−0.018	1

Appendix F: Determinants of T.E Thorough Random Effect Versus Fixed Effect

Variables	I	II
PAT	0.048 (2.06)**	0.053 (2.3)**
RDS	−0.005 (−0.67)	−0.005 (−0.74)
LIC	0.005 (0.16)	−0.015 (−0.46)
FOS	0.001 (0.3)	0.001 (0.3)
ID	−0.005 (−1.23)	−0.005 (−1.27)
Cons	0.804 (139.2)*	0.804 (414.61)*
sigma_u	0.017	0.041
sigma_e	0.106	0.106
rho	0.025	0.132
F test	–	F (9,4875) = 66.05
	Random effect	Fixed effect

Z values are in parenthesis. *, **Represent 1, 5 % level of significance

References

Aghion, P., Harris, C., Howitt, P., & Vickers, J. (2001). Competition, imitation and growth with step-by-step innovation. *The Review of Economic Studies, 68*(3), 467–492.

Aitken, B. J., & Harrison, A. E. (1999). Do domestic firms benefit from direct foreign investment? Evidence from Venezuela. *American Economic Review, 89*, 605–618.

Bartelsman, E., & Doms, M. (2000). Understanding productivity: Lessons from longitudinal micro data. *Journal of Economic Literature, 38*(3), 569–594.

Basant, R., & Fikkert, B. (1996). The effects of R&D, foreign technology purchase, and domestic and international spillovers on productivity in Indian firms. *The Review of Economics and Statistics, 78*(2), 187–199.

Bhandari, A. K., & Ray, S. C. (2012). Technical efficiency in the Indian textile industry: A non parametric analysis of firm-level data. *Bulletin of Economic Research, 64*(1), 109–124.

Bremmer, J., Alfons, G. J. M., Lansik, O., & Huirne, R. B. M. (2008). The impact of innovation, firm growth and perceptions on technical and scale efficiency. *Agricultural Economic Review, 9*(2), 65–85.

Cameron, A. C., & Trivedi, P. K. (2010). Micro econometric using stata (Revised ed.). Lakeway Drive, Texas: Stata Press Publication. 4905.

Chang, X., McLean, R. D., Zhang, B., & Zhang, W. (2013). *Patents and productivity growth: Evidence from global patent awards*. Available at http://papers.ssrn.com/sol3/papers.cfm?abstract_id=2371600

Charnes, A., Cooper, W. W., & Rhodes, E. (1978). Measuring the efficiency of decision-making units. *European journal of operational research, 3*, 429–444.

Coe, D. T., & Helpman, E. (1993). *International R&D spillovers*, NBER working paper series 4444. Available at http://www.nber.org/papers/w4444

Crepon, B., Duguet, E., & Mairesse, J. (1998). *Research, innovation and productivity: An econometric analysis at the firm level*, NBER working paper 6696. Available at http://www.nber.orgpapers/w6696

Deolalikar, A. B., & Roller, H. L. (1989). Patenting by manufacturing firms in India: Its production and impact. *The Journal of Industrial Economics, 37*(3), 303–314.

Diaz-Mayans, M. A., & Sanchez, R. (2014). *Innovation, exports and technical efficiency*, MPRA paper No 53230. Available at http://mpra.ub.uni-muenchen.de/53230/

Eaton, J., & Kortum, S. (1996). Trade in ideas: patenting and productivity in the OECD. *Journal of International Economics, 40*, 251–278.

Farrell, M. J. (1957). The measurement of productive efficiency. *Journal of the Royal Statistical Society, 120*(3), 253–290. Series A (General).

Fethi, M. D., Jackson, P. M., & Weyman-Jones, T. G. (2000). *Measuring the efficiency of European airlines: An application of DEA and Tobit analysis*, EPRU Discussion paper. Available at http://hdl.handle.net/2381/370

Gillen, D., & Lall, A. (1997). Developing measures of airport productivity and performance: An application of data envelopment analysis. *Transportation Research, 33*(4), 261–273.

Griffith, R., Huergo, E., Mairesse, J., & Peters, B. (2006). *Innovation and productivity across four European countries*, NBER working paper series 12722. Available at http://www.nber.org/papers/w12722

Griliches, Z. (1987). Patent statistics an economic indicator: A survey. In Z. Griliches (Ed.), *R&D and productivity: The econometric evidence*. University of Chicago press.

Griliches, Z., & Mairesse, J. (1984). Productivity and R&D at the firm level. In Z. V. Griliches (Ed.), *R&D, patents and productivity*. Massachusetts Avenue Cambridge, U.S.A: NBER Books.

Hall, B. H. (2011). *Innovation and productivity*, NBER working paper No. 17178. Available at http://www.nber.org/papers/w17178

Honore, B. E. (1992). Trimmed lad and least squares estimation of truncated and censored regression models with fixed effects. *Econometrica, 60*(3), 533–565.

Kamien, M. I., & Schwarts, N. L. (1982). *Market structure and innovation*. Cambridge: Cambridge University Press.

Kathuria, V., Raj, R. S. N., & Sen, K. (2013). Productivity measurement in Indian manufacturing: A comparison of alternative methods. *Journal of Quantitative Economics, 11*(1&2), 148–179.

Krohonen, P., Tainio, R., & Wallenius, J. (2001). Value efficiency analysis of academic research. *European Journal of Operational Research, 130*, 121–132.

Kumar, S. (2005). A decomposition of total productivity growth: A regional analysis of Indian industrial manufacturing growth. *International Journal of Productivity and Performance Management, 55*(3/4), 311–331.

Levinsohn, J., & Petrin, A. (2003). Estimating production functions using inputs to control for unobservables. *The Review of Economic Studies, 70*(2), 317–341.

Luoma, K., Jarvio, M. L., Suoniemi, I., & Hjerppe, R. T. (1996). Financial incentives and productive efficiency in Finnish health centers. *Health Economics, 5*, 435–445.

Mahadevan, R. (2001). Assessing the output and productivity growth of Malaysia's manufacturing sector. *Journal of Asian Economics, 12*, 587–597.
Mahadevan, R. (2003). To measure or not to measure total factor productivity growth? *Oxford Development Studies, 31*(3), 365–378.
Mansfield, E. (1984). R&D and innovation: Some empirical findings. In Z. V. Griliches (Ed.), *R&D, patents and productivity*. Massachusetts Avenue Cambridge, U.S.A: NBER Books.
Marjit, S., & Kar, S. (2009). A contemporary perspective on the informal labour market: Theory, policy and the Indian experience. *Economic and Political Weekly, 44*(14), 60–71.
Meeusen, W., & Van den Broeck, J. (1997). Efficiency estimation from Cobb-Douglas production functions with composed errors. *International Economic Review, 18*, 435–444.
Mokyr, J. (2010). The contribution of economic history to the study of innovation and technical change: 1750–1914. In B. H. Hall & N. Rossenberg (Eds.), *Handbook of the economics of innovation* (Vol. I). Oxford, U.K.
Raj, R. S. N. (2011). Structure, employment and productivity growth in the Indian unorganized manufacturing sector: An industry level analysis. *Singapore Economic Review, 56*, 349–376.
Raut, L. K. (1995). R&D spillover and productivity growth: Evidence from Indian private firms. *Journal of Development Economics, 48*, 1–23.
Ray, S. C. (2002). Did India's economic reforms improve efficiency and productivity? A non parametric analysis of the initial evidence from manufacturing. *Indian Economic Review, 37* (1), 23–57.
Sanchez, R., & Diaz, M. A. (2013). *Are large innovative firms more efficient?* MPRA paper No 44592. Available at http://mpra.ub.uni-muenchen.de/44592/
Sharma, C. (2010). Does productivity differ in domestic and foreign firms? Evidence from Indian machinery industry. *Indian Economics Review, 45*(1), 87–110.
Wang, E. C., & Huang, W. (2007). Relative efficiency of R&D activities: A cross-country study accounting for environmental factors in the DEA approach. *Research Policy, 36*, 260–273.

Business Group Affiliation and Innovation in Medium and High-Technology Industries in India

Surenderrao Komera, P.J. Jijo Lukose and Subash Sasidharan

1 Introduction

A distinct feature of many of the economies is the presence of business group (BG) firms. These group affiliated firms are known by various names such as Keiretsu in Japan, Chaebol in Korea, Jituanqiye in Taiwan, grupos economics in Latin America, business houses in India. Even though there are definitional variations, and these are legal entities, Khanna and Rivkin (2001) define business groups firms as a set of firms which, 'though legally independent, are bound together by a constellation of formal and informal ties, and are accustomed to taking coordinated action'. They act like a network and there is evidence that considerable resource sharing among affiliates is widely prevalent (Hsieh et al. 2010). Such groups are considered as a way of ameliorating the financing constraints of the affiliates through the creation of internal capital market. An interesting question emerging from the working of internal capital market is whether the group affiliation hinders or promotes innovative activities.

The existing studies have mainly focused on performance related aspect of group affiliation (Carney et al. 2011). There are very few studies that focus on the strategies (i.e. R&D) of the business group firms. According to Mahmood and Lee (2004) "*We do not know, however, whether or not there is any systematic relationship between group structure and technological innovation*". Therefore, there is a need to probe whether the superior financial performance of business group

S. Komera (✉)
Institute for Financial Management and Research, Chennai 600034, India
e-mail: surender@ifmr.ac.in

P.J.J. Lukose
Indian Institute of Management Kozhikode, Kozhikode 673570, India

S. Sasidharan
Indian Institute of Technology Madras, Chennai 600036, India

N.S. Siddharthan and K. Narayanan (eds.), *Technology*,
India Studies in Business and Economics,
DOI 10.1007/978-981-10-1684-4_4

affiliation stems from their distinct strategies like investments in R&D. In an influential study, Chang et al. (2006) discuss in detail how group affiliation fosters innovative efforts of emerging market firms. *First*, group affiliation provides valuable source of finance which otherwise is difficult to obtain from the market. Due to market imperfections, BGs in developing economies can overcome the problem of liquidity for investment in R&D activities (Belenzon and Berkovitz 2010; Hsieh et al. 2010). *Second*, the absence of proper intellectual property rights protection may prevent foreign firms from entering into technology collaboration with the developing economy firms. Due to the reputation problem, BGs are likely to honor the intellectual property rights and contract enforcement. Hence, foreign firms are likely to collaborate and share technology with group affiliates than stand-alone firms. *Third*, BGs can provide internal labor market for high skilled employees. *Finally*, group affiliation provides scope for knowledge spillover. Complementarity of the resources of the affiliates helps them in enhancing the innovativeness and improving the infrastructure.

Business groups are a unique feature of many of the emerging economies like China, Korea, India, and Taiwan. It is widely held that the presence of strong business groups in emerging economies results in more investment in innovative activities (Mahmood and Lee 2004). However, some studies argue that BG can hinder innovative activities by preventing new entrants and lowering technology accumulation. Given the importance of BGs in emerging economies; it is crucial to study their role in fostering innovation. The existing studies mainly focus on the factors determining business group formation (Maman 2002), performance effects (Khanna and Palepu 2000), and more recently innovation (Belenzon and Berkovitz 2010). Therefore, the main focus of this paper is to explore the relationship between group affiliation and innovation. This assumes greater relevance as the adoption of economic reform measures by economies like India ushered competitive challenges forcing BGs to innovate and maintain dominance (Aghion et al. 2005). Further, India's unique corporate sector setting provides an interesting case study as BG firms are considered as the main drivers of innovation and economic growth (Chakrabarti et al. 2008). Therefore, the empirical evidence in this study can provide key insights in understanding the strategies of the BG firms in a developing economy setting.

This study contributes to the literature in many ways. This is possibly one of the first studies using the patent data obtained from the Indian Patent Office (IPO) to analyze relationship between ownership structure and innovation. Even though, R&D expenditure may be widely used as an indicator of innovation, patenting activity reflects the quality and extent of firm innovation (Griliches 1990; Bernstein 2015). Though there are exceptions (Singh and Gaur 2013; Lodh et al. 2014), studies from the Indian context mainly focused on impact of financial factors and foreign ownership on innovation. The present study contributes to an emerging strand in the literature on the relationship between ownership structure and innovative activity of firms (Aghion et al. 2013; Bernstein 2015).

The reminder of the chapter is organized as follows. Section 2 discusses the relevant literature. Section 3 presents empirical model and data source. Section 4 presents the econometric results and discussion. Section 5 provides concluding remarks.

2 Related Literature

BG affiliation is often considered as an important factor in determining long- term investments and innovative efforts by firms in emerging markets. As discussed in the previous section, BG affiliation is expected to contribute more to innovation due to a number of reasons such as resource sharing among the group firms, presence of internal capital market, healthy competition among group affiliates, and possibility of sharing research outputs. Emerging markets are known for weaker institutional infrastructure and governance leading to limited resources for investment in R&D. Business groups that are understood to fill the institutional voids are expected to fund their affiliates' R&D activities by substituting/complementing external sources with their internal capital markets (Khanna and Palepu 1997; Carney et al. 2011). Some of the most important empirical studies examining this relationship in various institutional settings such as Italy, Korea, France, Taiwan etc. are summarized in Table 1. It is evident from the table that previous studies have employed both R&D expenditure and patent count as measures of innovation. Apart from using BG affiliation, some studies have used measures that capture group structure such as group diversification (Chang et al. 2006), firm's position within the group using corporate control-chain (Filatotchev et al. 2003). Further, most of the studies are based on the experience of the developed markets.

In India, BGs are organized through a central governance structure with several private and/or listed affiliated firms (Kedia et al. 2006). The BGs have differing control rights over these affiliates who are separate legal entities. An extensive literature on business groups has linked their existence to the weak institutional development in India (Khanna and Palepu 2000). The evolution of the BGs after the structural reforms in 1991 is often considered as a case study on how domestic businesses with financial muscle respond to the liberalization and globalization in an emerging economy (Sarkar 2010). Given the above background, it is interesting to examine how the BGs have fared in terms of embarking on innovative activities. Some studies including Singh and Gaur (2013), Komera et al. (2014) examined this relationship in the context of India. The results of these studies indicate a positive relationship between R&D intensity and group affiliation. Using an unbalanced panel of 395 listed firms, Lodh et al. (2014) find that affiliation with top 50 business groups increase innovation activities of family firms. Unlike Lodh et al. (2014), our study focuses on the patenting behavior of BGs and stand-alone firms directly rather than focusing on the family owned enterprises.

Table 1 Summary of empirical studies on BG affiliation and innovation

Authors	Type of data	Country	Dependent variable	Independent variable	Results
Filatotchev et al. (2003)	Firm level/2002 manufacturing firms	Italy/1992–94	R&D expenses/employees	Group dummy, position within the group (head or controlled)	Group affiliation positively associated with R&D intensity
Blanchard et al. (2005)	Firm level and group level/3100 firm affiliates	France/1994–98	Value added firm R&D	Group R&D	(1) Within group R&D has an impact on the productivity of affiliates (2) evidence of sharing of R&D benefits among group affiliates
Chang et al. (2006)	Firm level and group level/sample from the largest 500 firms	Korea Taiwan/1991–99	US patents Group dummy	Group diversification	(1) Affiliation benefited firm innovation in Korea, but not in Taiwan (2) Diversification at the group level negatively affected affiliate firms' innovation performance in Taiwan, where capital markets were more developed than they were in Korea
Kim et al. (2008)	Firm level/253 firms	Korea/1998–2003	Ratio of R&D expenditures to total sales	Group dummy	Do not find evidence of group affiliation influencing R&D investments
Cefis et al. (2009)	Firm level/3696 Italian firms with a positive R&D budget	Italy/1992	R&D expenses on product and process innovation	Group dummy	Firms that belong to a group invest significantly more into aggregate R&D than independent firms
Singh and Gaur (2013)	Firm level/16,337 firm-year observations	India 2002–2009	R&D expenses over total sales	Group affiliation dummy	Positive effect of group affiliation on R&D intensity
Belenzon and Berkovitz (2010)	Firm level/11,645 firm of which 5683 are affiliated	Europe/1995–2004	EPO and US patents	Group dummy, group diversification, industry dependence on external finance,	(1) Group affiliates patent more than standalones or affiliates of small groups (2) Group-innovation relation is stronger in industries that rely more on external

(continued)

Table 1 (continued)

Authors	Type of data	Country	Dependent variable	Independent variable	Results
				research similarity index of affiliated firms	finance and in groups with more-diversified capital sources
Hsieh et al. (2010)	Firm level and group level/4170 firms of which 715 belonging to a group	Taiwan/2001–2003	Granted patent applications	Group dummy, group diversification, industry dependence on external finance	(1) Business groups foster innovation of affiliated firms (2) Group-innovation relation is stronger in industries that rely more on external finance (3) Group diversification is positively related to firm innovativeness
Guzzini and Iacobucci (2014a, b)	Firm level/4000 firms	Italy/2001–2003	Dummy variable for R&D expenses, R&D intensity	Group dummy, group diversification	Positive association between R&D investment and group affiliation
Guzzini and Iacobucci (2014a, b)	SMEs/3446 firms	Italy/2001–2003	R&D propensity and R&D intensity	Bottom group dummy, head/intermediate dummy	Positive association between R&D propensity and head/intermediate affiliate firms
Lodh et al. (2014)	395 listed firms	2001–2008 India	R&D intensity and patent productivity	Family ownership interacted with business group	Affiliating with top 50 business groups increases the innovation activities

Source Compiled from Guzzini and Iacobucci (2014a, b)

3 Data Source and Methods

3.1 *Sample*

The sample firms consist of medium and high-tech firms based on the OECD technology intensity classification. The classification includes mainly Pharmaceuticals, Chemicals, Automobiles, Electronics, and Machinery. We obtained 7690 firm year observations of National Stock Exchange (NSE) listed firms with financial information during the period 2001–10. Patent information is available for 1914 firm year observations pertaining to 232 firms. 748 Firms with no patent data contributed to the remaining 5776 firm year observations. Average number of observations per firm for both categories is 8. As a measure of firm heterogeneity, we estimated pre-sample mean during the period of 1995–2000.

3.2 *Variables*

3.2.1 Dependent Variable

We considered patents as measure of innovative efforts by sample firms. Similar to US Patent and Trademark office (USPTO) and European Patent and Trademark office (EPTO), their Indian counterpart Indian Patent office (IPO) under the Controller General of Design, Trademark and Patent (CGPDTM) provides information on patents obtained by firms in India. We used the patent data obtained from the electronic search engine of IPO.[1]

3.2.2 Independent Variables

Our data source for independent variables is the PROWESS database maintained by Center for Monitoring Indian Economy. The variables used in the empirical analysis are drawn from the annual report of the firms. This database was previously employed by many firm level studies for analyzing the performance of business group firms (Khanna and Palepu 2000; Chittoor et al. 2014) and analyzing R&D investment by firms (Singh and Gaur 2013; Lodh et al. 2014).

BG affiliation is the key explanatory variable in various econometric specifications given in the next sub-section. We consider the Center for Monitoring Indian Economy's definition in classifying the sample firms into business group affiliated or stand-alone firms. Similar criterion has been employed previously (Singh and Gaur 2013; Lodh et al. 2014). We control for profitability, leverage, firm size, age, and time effects in all our specifications. Following earlier studies, *profitability*,

[1]The same data set was previously used by Ambrammal and Sharma (2015).

defined as sum of profit after tax, depreciation, amortization, and R&D expenses (scaled by total assets) is considered as the proxy for firms' financial constraints. *Leverage* is defined as book value of debt divided by total assets. We expect a negative relationship between leverage and innovation as technologically intensive firms prefers equity financing for new investments (Gugler 2001). *Firm size* is measured as natural logarithm of total assets. Previous studies based on India have reported a non-linear relationship between firm size and R&D (Siddharthan 1988). We control for *Age*, computed as the number of years since incorporation. Experienced firms are expected to have better expertise in exploiting the benefits of R&D efforts.

3.3 Econometric Model and Estimation

The basic structure of the patent data poses many challenges for the econometric analysis. Following Belenzon and Berkovitz (2010), the relationship between innovations (patent count) of firm i in period t and BG affiliation can be expressed as

$$E(P_{it}|X_{it}) = \exp[\alpha\, Group_i + \beta \mathbf{Z}_{it} + \tau_t] \quad (1)$$

where $Group_i$ is a dummy variable with value 1 if the firm belongs to business group and 0 otherwise. $\mathbf{Z}_{it}$ represents the vector of control variables. τ_t is the standard time dummy. In some of the specifications, we used R&D expenditure as an additional control variable to account for the impact of BGs in raising R&D (Singh and Gaur 2013; Komera et al. 2014). In these specifications, α capture innovative performance of business firms compared to stand-alone firms. The above basic specification with long-link transformation accounts for count nature of the data. Following Aghion et al. (2013), we report two sets of estimators using poisson and negative binominal model. The poisson regression model captures the count nature of the data by restricting conditional mean of the dependent variable as positive. But it makes a restrictive assumption about the equality of conditional mean and conditional variance of the dependent variable. The negative binomial model is a more flexible functional form that can deal with the problem of over-dispersion, typically observed in count data (Cameron and Trivedi 2005).

Firm's selection into BG in India is not endogenous; therefore the self-selection bias is unlikely to affect our estimates. Another potential bias may arise from the persistence in the innovative activity among the sample firms. The unobserved heterogeneity at the beginning of the sample period, may lead to biased estimates. Introducing firm fixed effects by way of pre-sample means (Blundell et al. 1999) mitigates this problem. We used average of patent counts up to five years prior to the sample period. However, we believe that given the absence of group selection bias, the model without this additional variable could capture innovative performance associated with BGs.

4 Results and Discussion

4.1 Summary Statistics

Table 2 presents the summary statistics of the variables included in the empirical analysis by classifying the sample firms into BG affiliated firms and stand-alone firms. All the variables included are winsorized at one and 99 ‰ to eliminate the effect of outliers. The average of the patent count for the entire sample firms is 0.3 and average R&D intensity is 0.85 %. This average for the medium and high-tech firms in India is comparable to the average EPTO patent values (0.28) for European manufacturing firms (Belenzon and Berkowitz 2010). On average, a BG affiliated firm has better innovation output by way of patent count (0.35) compared to stand-alone firm (0.27). Similarly, the R&D intensity of BG affiliates is significantly higher than stand-alone counterparts. Moreover, BGs are larger, profitable, and more leveraged.

4.2 Group Affiliation and Innovation

After examining the basic characteristic of the sample firms, we estimate Eq. 1 using both Poisson and Negative binomial model. Table 3 presents the basic results. Columns 1 and 2 report the poisson regressions; columns 3–6 provide the

Table 2 Summary statistics

Variable	All firms		Business group firms		Stand-alone firms		*p*-value
	Mean	Median	Mean	Median	Mean	Median	
Patent count	0.30	0.00	0.35	0.00	0.27	0.00	0.027**
Total assets	516.89	97.37	765.54	222.28	352.73	50.05	0.000***
R&D to total assets (%)	0.85	0.05	0.99	0.22	0.75	0.00	0.003***
Cash flow to total assets	0.11	0.09	0.11	0.10	0.10	0.09	0.003***
Leverage	0.28	0.27	0.29	0.29	0.27	0.26	0.000***
Observations	7690		3058		4632		

Table provides mean and median values of the sample firms classified based on ownership classification as BG affiliated firms and stand-alone firms. All the ratio variables included are winsorized at one as well as at 99 ‰ to eliminate the effect of extreme values. The reporting figures are the corresponding averages. The differences between BG affiliated firms and stand-alone firms are tested for their statistical significance using the t-statistics. The corresponding *p*-values are reported. *LEV* (Leverage) is defined as the ratio of total borrowing to total assets; *Size* is the natural logarithm of total assets (in rupees crores); *CF* (Cash flow) is the sum of profit after tax, depreciation, amortization, and R&D expenditure divided by total assets; *AGE* of firms in years since incorporation; *R&D intensity* is measured by the ratio between R&D expenditure and total assets in percentage

*** and ** Refer to 1 and 5 % significance level

Table 3 Innovation and business group affiliation

	Poisson		Negative binomial			
	1	2	3	4	5	6
BG dummy	−0.274***	−0.309***	−0.461***	−0.487***	−0.377***	−0.398***
	(0.000)	(0.000)	(0.001)	(0.000)	(0.005)	(0.003)
R&D_TA		0.101***		0.181***		0.215***
		(0.000)		(0.000)		(0.000)
Cash flow	5.339***	3.906***	5.242***	2.848***	5.463***	2.411***
	(0.000)	(0.000)	(0.000)	(0.001)	(0.000)	(0.004)
Leverage	−0.353**	−0.627***	−1.583***	−2.306***	−0.356	−1.101***
	(0.034)	(0.000)	(0.000)	(0.000)	(0.380)	(0.009)
Age	0.125***	0.169***	−0.002	0.006	−0.163*	−0.165*
	(0.002)	(0.000)	(0.986)	(0.950)	(0.076)	(0.068)
Size	0.529***	0.499***	0.898***	0.875***	0.793***	0.759***
	(0.000)	(0.000)	(0.000)	(0.000)	(0.000)	(0.000)
Pre-sample fixed effect	0.032***	0.033***			0.151***	0.159***
	(0.000)	(0.000)			(0.000)	(0.000)
Time	Yes	Yes	Yes	Yes	Yes	Yes
Log pseudo likelihood	−3813.0	−3754.9	−2267.2	−2251.3	−2201.7	−2177.8
Test of over-dispersion (LR test *p*-value)			0.000	0.000	0.000	0.000
Observations	6418	6418	6418	6418	6418	6418

Table presents the findings from estimating Eq. 1 using poisson and negative binomial regression estimators. We also present the *p*-values of the LR test statistic which validates the use of negative binomial regression estimator. The numbers in the parenthesis are corresponding *p*-values. The description of the variables is provided in the notes to Table 2

***, **, and * Refer to 1, 5 and 10 % significance level

estimates using negative binomial regressions. We have estimated the same specification using an OLS regression with natural logarithm of patents as the dependent variable. As the three estimates provide similar results, for brevity we are not reporting the OLS results. Further, due to the over-dispersion of the data, we focus on the results obtained from the negative binomial model. The coefficients of the control variables are consistent with the existing literature. We find that R&D intensity, age, size, and profitability positively influence patenting activity. On the other hand, leverage has a negative impact on patenting. Our results also indicate considerable persistence in patenting activity, as the coefficient of the pre-sample mean is positive and significant. The key finding of the study is that after controlling for firm characteristics, coefficients of BG affiliation lie within a range of −0.38 to −0.49 indicating a lower innovation output. The negative effect of BG affiliation on innovation may be due to: *first*, the sample firms in this study belong to medium and high-tech firms where the R&D intensity (0.85 %) is almost four times as compared to the overall manufacturing sector (0.23 %). Therefore, we caution that the dynamics of the patenting activity in technologically intensive sector may not be generalizable to the entire manufacturing. *Second*, There is ample evidence with respect to the Indian BGs that the frequency with which they bring products and technologies to the market may hinder them from undertaking time consuming patenting (Lodha et al. 2014). Further, most of these innovations are incremental that it may be difficult to obtain patents. *Finally*, if BGs manage all patent related activity through a single firm and share the resources with other affiliates, the result may show a downward bias for BG affiliates if the single firm is not coming under medium and high technology category (Belenzon and Berkovitz 2010).

4.3 Evolution of Effect of BG on Innovation

One strand in the literature often argues that the benefits of business group affiliation decline as the institutional environment in emerging markets like India improves over a period of time. The above 'institutional voids argument' posits that BGs in emerging markets substitute underdeveloped capital market, labor market, etc., organizing the same internally. Therefore, we should expect a declining trend in the relationship between BG and innovation over a period of time. As a subsequent step, we examine the evolution of patenting activity during the sample period using the following model

$$E(P_{it}|X_{it}) = \exp[\alpha\, Group_i + \beta \mathbf{Z}_{it} + \gamma\, TimeClock_t + \tau_t] \tag{2}$$

$$E(P_{it}|X_{it}) = \exp[\alpha\, Group_i + \beta \mathbf{Z}_{it} + \gamma_1 TimeClock_t + \gamma_2 Group_i * TimeClock_t + \tau_t] \tag{3}$$

Table 4 Evolution of effect of BG affiliation on innovation

	Negative binomial	
	1	2
BG dummy	−0.398***	−0.506*
	(0.003)	(0.082)
R&D_TA	0.215***	0.215***
	(0.000)	(0.000)
Cash flow	2.411***	2.416***
	(0.004)	(0.004)
Leverage	−1.101***	−1.102***
	(0.009)	(0.009)
Age	−0.165*	−0.164*
	(0.068)	(0.070)
Size	0.759***	0.757***
	(0.000)	(0.000)
Pre-sample fixed effect	0.159***	0.159***
	(0.000)	(0.000)
TimeClock	−0.828***	−0.842***
	(0.000)	(0.000)
TimeClock * BG dummy		0.025
		(0.677)
Time	Yes	Yes
Log pseudo likelihood	−2177.8	−2177.7
Test of over-dispersion (LR test *p*-value)	0.000	0.000
Observations	6418	6418

Table presents the findings from estimating Eq. 2 and e using negative binomial regression estimator. We also present the *p*-values of the LR test statistic which validates the use of negative binomial regression estimator. The numbers in the parenthesis are corresponding *p*-values. The description of the variables is provided in the notes to Table 2
*** and * Refer to 1, 5 and 10 % significance level

where '*TimeClock*' refers to the variable we constructed to measure the time trend. Table 4 presents the empirical findings of Eqs. 2 and 3 respectively. The basic results remains intact even after controlling for time trend. Results indicate that sample firms show a declining trend in the patenting activity during the study period. Though, the interaction dummy of BG with time clock is positive, it is insignificant.

4.4 Robustness Check: Propensity Score Matching

As Table 2 highlights differences between BG affiliates and standalone firms, we used an alternate empirical method (matching estimator) to check the robustness of

Table 5 Difference in patent count between BG affiliated and stand-alone firms

Dependent variable	Treatment variable	Coefficient	Z ($P > \|Z\|$)	No of observations
Patent count	BG versus standalone	−0.0363	−0.92 (0.321)	6418

The matching estimator has been the average treatment effect. The business group affiliated firms are matching with the nearest neighbor stand-alone matching firm across the following variables: size, age, leverage, pre-sample mean and cash flow. Number of matches = 2. The above estimates are heteroskedastic consistent

the results. Table 5 shows the result of average treatment effect (ATT) using the bias corrected matching estimator (Abadie et al. 2004). Specifically, the control firms are the closest matches to the BG firms based on a set of continuous variables such as size, age, leverage, pre-sample mean and cash-flow. The mean of the differences in patents between BG and standalone firm is −0.036. The test statistic shows that these differences are insignificant. This casts doubt on the significance of the evidence reported in the previous section. Nevertheless, still the mean difference shows that BG affiliated firms are less innovative compared to a similar stand-alone firms matched in terms of their firm specific characteristics.

5 Conclusions

The present study examines whether business group affiliation fosters innovation in medium and high technology industries in India, using a unique dataset of patents obtained from Indian patent office. Our results indicate that BG affiliates show more R&D intensity and obtained more patents compared to stand-alone firms during the study period. However, controlling for firm specific characteristics, business group affiliation seems to hinder innovation in technology intensive sectors. A number of plausible reasons can be attributed to the above results and are discussed in the previous section. Further, we report persistence in patenting activity among sample firms. Innovation productivity is positively related to firm size, profitability, and age. Leverage is negatively related to patenting activity of the sample firms.

In spite of the careful attempt, the results reported may be biased if there are systematic biases in patenting activity between BG affiliates and stand-alone firms. Various studies highlight the presence of a parent firm (also known as headquarters) with each BG dealing with legal, managerial and related aspects pertaining to the group. If patenting is routed through headquarters that are not in the medium and high technology industries, there is a possibility of bias in our findings. A possible extension of the study by including firms belonging to entire manufacturing sector can address this concern.

References

Abadie, A., Drukker, D., Herr, J. L., & Imbens, G. W. (2004). Implementing matching estimators for average treatment effects in Stata. *The Stata Journal, 4*(3), 290–311.

Aghion, P., Burgess, R., Reddding, S., & Zillibotti, F. (2005). Entry liberalization and inequality in industrial performance. *Journal of European Economic Association, 3*, 291–302.

Aghion, P., Reenen, J. V., & Zingales, L. (2013). Innovation and institutional ownership. *American Economic Review, 103*(1), 277–304.

Ambrammal, S. K., & Sharma, R. (2015). Impact of patenting on firms' performance: An empirical investigation based on manufacturing firms in India. *Economics of Innovation and New Technology*. doi:10.1080/10438599.2015.1043767

Belenzon, S., & Berkovitz, T. (2010). Innovation in business groups. *Management Science, 56*(3), 519–535.

Bernstein, S. (2015). Does going public affect innovation? *Journal of Finance, 70*, 1365–1403.

Blanchard, P., Huiban, J. P., & Sevestre, P. (2005). R&D and productivity in corporate groups: An empirical investigation using a panel of French firms. *Annalesd'économieet de Statistique, 79* (80), 461–485.

Blundell, R., Griffith, R., & Van Reenen, J. (1999). Market share, market value and innovation in a panel of British manufacturing firms. *The Review of Economic Studies, 66*, 529–554.

Cameron, A. C., & Trivedi, P. K. (2005). *Microeconometrics: Methods and applications*. New York: Cambridge University Press.

Carney, M., Gedajlovic, E. R., Heugens, P. P. M. A. R., Essen, M., & Oosterhout, J. H. (2011). Business group affiliation, performance, context, and strategy: A meta-analysis. *Academy of Management Journal, 54*(3), 437–460.

Cefis, E., Rosenkranz, S., & Weitzel, U. (2009). Effects of coordinated strategies on product and process R&D. *Journal of Economics, 96*(3), 193–222.

Chakrabarti, R., Megginson, W., & Yadav, P. K. (2008). Corporate governance in India. *Journal of Applied Corporate Finance, 20*, 59–72.

Chang, S. J., Chung, C. N., & Mahmood, I. P. (2006). When and how does business group affiliation promote firm innovation? A tale of two emerging economies. *Organization Science, 17*(5), 637–656.

Chittoor, R., Kale, P., & Puranam, P. (2014) Business groups in developing capital markets: Towards a complementary perspective. Strategic Management Journal. doi:10.1002/smj.2287

Filatotchev, I., Piga, C., & Dyomina, N. (2003). Network positioning and R&D activity: A study of Italian groups. *R&D Management, 33*(1), 37–48.

Griliches, Z. (1990). Patents statistics as economic indicators: A survey. *Journal of Economic Literature, 28*(4), 1661–1707.

Gugler, K. (2001). *Corporate governance and economic performance*. Oxford: Oxford University Press.

Guzzini, E., & Iacobucci, D. (2014a). Business group affiliation and R&D. *Industry and Innovaton, 21*(1), 20–42.

Guzzini, E., & Iacobucci, D. (2014b). Ownership as R&D incentive in business groups. *Small Business Economics, 43*, 119–135.

Hsieh, T., Yeh, R., & Chen, Y. (2010). Business group characteristics and affiliated firm innovation: The case of Taiwan. *Industrial Marketing Management, 39*(4), 560–570.

Kedia, B. L., Mukherjee, D., & Lahiri, S. (2006). Indian business groups: Evolution and transformation. *Asia Pacific Journal of Management, 23*, 559–577.

Khanna, T., & Palepu, K. (1997). Why focused strategies may be wrong for emerging markets? *Harvard Business Review, 75*(4), 41–51.

Khanna, T., & Palepu, K. (2000). Is group affiliation profitable in emerging markets? An analysis of diversified Indian business groups. *Journal of Finance, 55*(2), 867–891.

Khanna, T., & Rivkin, J. W. (2001). Estimating the performance effects of business groups in emerging markets. *Strategic Management Journal, 22*(1), 45–74.

Kim, H., Kim, H., & Lee, P. M. (2008). Ownership structure and the relationship between financial slack and R&D investments: Evidence from Korean firms. *Organization Science, 19*(3), 404–418

Komera, S. R., Lukose, J., & Sasidharan, S. (2014, October 27–29). Does business group affiliation encourage R&D activities? in *Evidence from India, presented at the Ninth Annual Conference of the Forum for Global Knowledge Sharing.*

Lodh, S., Nandy, M., & Chen, J. (2014). Innovation and family ownership: Empirical evidence from India. *Corporate governance: An international review, 22*, 4–23.

Mahmood, I. P., & Lee, C. Y. (2004). Business groups: Entry barrier-innovation debate revisited. *Journal of Economic Behavior & Organization, 54*, 513–531.

Maman, D. (2002). The emergence of business groups: Israel and South Korea compared. *Organization Studies, 23*(5), 737–758.

Sarkar, J. (2010). Business groups in India. In A. M. Colpan, T. Hikino, & J. R. Lincoln (Eds.), *The Oxford handbook of business groups*. Oxford University Press.

Siddharthan, N. S. (1988). In-house R&D, imported technology and firm size: Lessons from Indian experience. *Developing Economies, 26*, 212–221.

Singh, D., & Gaur, A. S. (2013). Governance structure, innovation and internationalization: Evidence from India. *Journal of International Management, 19*, 300–309.

Foreign Multinationals and Domestic Enterprises: Comparison of Their Technological and Other Characteristics in the Indian Machinery Industry

Pradeep Kumar Keshari

1 Introduction

The issue of divergence in the characteristics of FMEs and DEs has considerable importance for the policy and decision makers in the host developing countries and for the researchers interested in the study of multinational enterprises (MNEs), a major vehicle of FDI. If FMEs and DEs do not differ significantly, efforts to attract FDI and the study of MNEs as a separate field of enquiry would be largely futile. Main reasons for the recent interest in this topic is that the governments' world over are promoting FDI and associated activities of MNEs in their respective countries. The first decade of the 2000s have particularly witnessed sharp increase in the number of regulatory changes favouring FDI and international investment agreements (e.g. bilateral investment and double tax avoidance treaties) in developing and developed countries alike (UNCTAD 2014: Chap. 3). Some countries even offer concessions and incentives (e.g. tax rebates and tax holidays, assured return on investment, government guarantees) for promoting FDI (UNCTAD 2014: Chap. 3).

This is a revised version of the paper presented in the FGKS' IX Annual International Conference on Technology: Corporate and Social Dimensions held during October 27–29, 2014 at National Institute of Advanced Studies, Bengaluru. I also authored an earlier version of this paper which appeared as MPRA Paper: *Comparative performance of foreign affiliates and domestic firms in the Indian machinery industry,* https://mpra.ub.uni-muenchen.de/33076/.

The author has greatly benefitted from the suggestions for improvements received in the conference and afterward. The author gratefully acknowledges Prof. N.S. Siddharthan, MSE, Chennai and Prof Pravin Jha, JNU, New Delhi for the encouragements and comments received on the earlier drafts of this paper. The views expressed in the paper are entirely personal and do not belong to the IDBI Bank Ltd. in which the author works.

P.K. Keshari (✉)
IDBI Bank Limited, IDBI House, 44, Shakespeare Sarani, Kolkata 700017, India
e-mail: p.keshari@idbi.co.in

N.S. Siddharthan and K. Narayanan (eds.), *Technology*,
India Studies in Business and Economics,
DOI 10.1007/978-981-10-1684-4_5

The efforts undertaken by Government of India (GoI) since the onset of economic reforms in 1991 in the form of deregulation of industrial, trade and FDI policies, substantial liberalization of foreign exchange control regime and removal of trade related investment measures (TRIMs)[1] have all created level playing field for FMEs *vis a vis* DEs. Besides, GoI has adopted stronger intellectual property rights (IPR) regime as per trade related intellectual property rights (TRIPs) agreement administered by World Trade Organisation (WTO) for protecting the interest of foreign investors. Further, it has also entered bilateral investment treaties with increasing number of countries. GoI has of late been permitting 100 percent foreign equity participation in manufacturing firms, except in MSE sector, on an automatic basis with a view to "supplement domestic capital, technology and skills for accelerating economic growth" (GoI 2013: p. 5). In the above background, it is important to understand the nature of differences between FMEs and DEs and the effects foreign ownership have in the Indian economy, particularly on its domestic industrial sector.

FDI being an important source of stable long term equity capital and critical firm-specific assets (FSAs), which are mostly technological and managerial in nature, is viewed beneficial to the developing countries on the basis of an important strand of theoretical literature and some empirical findings that FMEs are superior to DEs in terms of their holding of FSAs and several aspects of performance derived from these FSAs such as efficiency and exports (Dunning 2000; Kobrin 2005). Therefore, locating more FMEs in the host developing economies may lead to direct benefits on account of increased number of firms with superior FSAs and performance. Besides, the presence of FMEs in the host developing economies may indirectly cause benefits to DEs through horizontal and vertical linkages and by improving their efficiency level and export performance through increased competition and knowledge spillovers (Keshari 2012, 2013; Smeets 2008).

The extant empirical literature, however, suggests that the differences in the characteristics of FMEs and DEs are: (a) contextual that is country, industry and FDI specific and (b) differences between FMEs and uni-national DEs do exist but not between FMEs and multinational DEs (Lall and Narula 2004; Jungnickel 2002; Bellak 2004). In view of substantially increased attractiveness of India for FDI[2] in recent years, existence of level playing field between FMEs and DEs and paucity of firm-level and industry-specific studies in the Indian context,[3] this study attempts to identify major firm-specific characteristics which could discriminate between FMEs and DEs or enable a firm to fall in the category of FMEs or DEs in an industry. The major aspects of characteristics covered in this study are: nature of technology (in-house R&D, choice of technique, import of embodied and disembodied

[1]Major TRIMs in past included fulfillment of technology, local content, dividend balancing and export obligations on the part of foreign firms with FDI.

[2]India became the second most attractive destination (next to China) among MNEs for FDI in terms of A.T. Kearney's 2007 FDI Confidence Index (Global Business Policy Council 2008).

[3]To the best of my knowledge, there is only one firm level study in the recent periods comparing a few aspects of performance and conducts that is by Ray and Rahman (2006).

technology), product differentiation, firm size, age, financial leverage, export intensity, efficiency in the use of inputs of production and gross profit margin.

Rest of the study is organised as follows. Section 2 presents the analytical framework and reviews the relevant empirical literature. Section 3 forwards arguments, formulates hypotheses on individual aspects of discriminating characteristics of FMEs and DEs and discusses definition and measurement of variables. Section 4 explains the reasons for selecting IMI and period of study, data sources and major characteristics of the sample. Section 5 explains the statistical method and econometric models used in the study. Section 6 analyses, discuses and compares the results obtained from the use of univariate method, LDA and the estimation of probit and logit models. Section 7 presents the conclusions and recommendations of the study.

2 Analytical Framework and Empirical Literature

2.1 Analytical Framework

Analytical framework for the empirical analysis in this paper mainly follows a mix of the transaction cost/internalization (TCI) theory as explained in Hennart (2007) and eclectic approach of FDI as discussed in Dunning (2000). This literature along with additional literature on the subject suggests the following factors to be generally important in creating the major differences in the characteristics of FMEs and DEs. First, FMEs have privileged access to two categories of superior FSAs. The first category is named as *explicit and observable assets* that may *inter alia* include machinery and equipment, intellectual property rights and skilled labour. These assets can be bought or acquired by DEs from FMEs and thereby the latter may lose competitive advantage derived from these assets in a short span of time. However, the FMEs can maintain their *competitive advantage* based on *tacit and unobservable* FSA for much longer period (e.g. trade secrets, know-why, R&D capabilities, organizational and management practices, routines and culture, etc.).

Secondly, FMEs may be more flexible and aggressive in utilising the FSAs as they are less hindered by the inertia of the local system. Thirdly, by combining location-specific advantages and working in the institutional set up and policy environment applicable to a host country, FMEs may develop their unique set of advantages by enhancing and modifying the FSAs originally received from their parents' network. The institutional perspective of business strategy emphasizes that the resource endowment of the host economy and its institutional framework moderate the characteristics of FMEs, facilitates the development of their resources and capabilities and even generate new capabilities and new markets opportunities, especially in the emerging economies (Meyer et al. 2009). Rugman and Verbeke (2001) argue that export from a particular FME may arise from affiliate specific regional advantages that are grounded in FSA acquired from both the parent and host country location-bound advantages.

The eclectic theory of FDI emphasises that the FMEs and DEs may differ in terms of their competitive advantages based on the ownership or access to *monopolistic advantages*, possession of a bundle of scarce, unique and sustainable *resources and capabilities* and *competence* to identify, evaluate, and harness resources and capabilities from throughout the world and to integrate them with their existing resources and capabilities (Dunning 2000). This literature also points out that the competitive advantages of FMEs over DEs are partly generic and partly context specific (Ibid).

2.2 Empirical Literature

In the last two decades, there has been a growth in empirical literature on relative performance of FMEs and DEs in the manufacturing sector of developed and developing economies. These studies have mostly used firm-level data and econometric methods. There have also been a few important studies surveying relatively recent literature on the subject. Jungnickel's (2002) edited volume of studies compare the behaviour of FMEs and DEs in a number of European countries. Bellak's (2004) survey, based on the 54 studies mainly using firm-level data in panel framework, compares the various aspects of performance chiefly for the industries based in the developed countries. The research papers in Jungnickel's (2002) edited volume (e.g. Bellak and Pfaffermayr 2002) predominately address both the theoretical and methodological issues associated with comparison between FMEs and DEs. They also empirically test the differences between FMEs and DEs in terms of selected indicators, such as productivity, wages and R&D.

These papers arrived at the following major conclusions: First, the real difference in behaviour and performance lies between FMEs and uni-national DEs and not between FMEs and multinational DEs (Jungnickel 2002; Bellak 2004). Second, the superior economic performance of FMEs over DEs is observed in the areas of productivity, technology, wages, skills and growth rates but mixed results in case of profitability (Bellak 2004). Thirdly, the comparison between FMEs and DEs is inherently context specific, and hence there are different finding in different countries, industries, etc. Notably, the performance gaps disappear when firm and industry characteristics are controlled in a regression equation (Jungnickel 2002; Bellak 2004).

There is no published survey of empirical literature available in the context of developing countries for the recent decades.[4] An unpublished Ph.D. thesis containing survey of firm-level empirical studies conducted during two decades of

[4]There has been two old surveys of literature by Jenkins (1989, 1990) focusing on the developing countries' experience in the 1970s.

1990–2010 pertaining to the manufacturing sector of developing countries using econometric techniques suggest that in majority of the cases: (i) the scholars focus on one aspect of firms' characteristics at a time in a study, (ii) FMEs, as compared to DEs, are larger in size, more capital intensive, not more R&D intensive, more export oriented and spend more on import of intermediate goods and disembodied technology, (iii) FMEs are more productive/efficient but not necessarily more profitable (Keshari 2011: Chap. 5, pp. 88–148).

In the case of Indian manufacturing sector, only two empirical studies have examined the issue of differences in the several characteristics of FMEs and DEs simultaneously by applying multivariate LDA technique. The pioneering research of Kumar (1990: Chap. 2) reveals that the FMEs, as compared to DEs, are more vertically integrated; operate at larger scales; employ more skilled personnel; earn higher profit margin; and have product differentiation advantage due to possession of higher amount of intangible assets. This study, however, is dated and uses aggregated firm-level data for an ownership category in an industry. In such a study, the use of firm-level (or sometimes plant level) data is considered appropriate (Bellak and Pfaffermayr 2002).

At firm level, Ray and Rahman (2006) evaluate the discriminating conducts of foreign and local enterprises mainly in terms of innovative activities and in establishing linkages with the domestic (or foreign sector) sector. The study uses a stratified random sample of 338 firms, each one with at least Rs. 40 crore of annual sales turnovers for the year 1997/98, belonging to the Indian chemical, electronics and transport equipment industries. The findings of this study suggest that: (a) FMEs spend more on import of disembodied technologies than DEs; (b) they however do not significantly differ in terms of R&D intensity, indicating that FMEs do not make efforts to adapt their technologies to the Indian condition; (c) FMEs foster backward horizontal linkages with local suppliers of final goods but make less efforts to develop backward vertical linkages. Although Ray and Rahman's (2006) is a firm-level study, it uses cross-section data, excludes performance aspects and does not include IMI in the scope of their study. Moreover, they have not estimated probabilistic models (probit/logit) which would have enabled them to control industry or sub-industry level influences on the categorical dependent variable capturing the foreign or domestic ownership.

3 Variables and Hypotheses

The following sub-sections discuss a priori arguments and the empirical findings pertaining to various firm-specific characteristics of FMEs and DEs. These characteristics are divided into two categories namely, technological and others.

3.1 Technological

3.1.1 Choice of Technique (CAPI)

The choice of technique of production used by a firm in an industry is generally captured by its capital intensity. Theoretically, all firms belonging to an industry, by reasons of common technology, are expected to operate with the same level of capital intensity. However, the capital intensity of FMEs may be higher than that of DEs for the following reasons. First, DEs economize on use of capital (than labour) in developing countries because they generally face higher cost in raising capital (than FMEs) in the financial market. Secondly, FMEs may have affinity towards more capital-intensive segments of an industry as they are mostly affiliated to the firms headquartered in the developed countries which have comparative advantage in producing capital-intensive goods. Based on a survey of a number of empirical studies pertaining to the developing countries mainly for the decade of 1970s, Jenkins (1990) concludes that when local and foreign firms are often in direct competition, producing similar products at similar scale of output, both ownership groups tend to choose production techniques having similar capital intensities. There are not many studies examining the issue of choice of techniques for the recent periods. A few studies, Ramstetter (1999a) for Thailand and other East Asian countries and Ngoc and Ramstetter (2004) for Vietnam, suggest that FMEs are relatively more capital intensive than DEs.

3.1.2 Research and Development Intensity (RDI)

Traditionally, R&D activities have been centralized in the headquarters of the MNEs located in the developed countries for the following reasons. First, FMEs have privileged access to the stock of technology and R&D laboratories located at their respective headquarters. Secondly, centralization of R&D activities at home location enables maintenance of secrecy, prevents leakage of FSAs to the competitors, minimize coordination costs and principal-agent problem.

Hence, FMEs have been undertaking only *asset exploiting* R&D activities in the host countries which involves minor expenditure for absorbing the technology, adapting intermediate goods obtained from the MNE systems and customization of final products to the peculiarities of local demand, regulations and standards of the host countries. Since the 1990s, MNEs have been shifting R&D activities from their respective headquarters to the locations of their FMEs in select developing countries, including India and China (Siddharthan 2009). It is also reported that the FMEs are complementing the traditional *asset exploiting* R&D activities with *asset augmenting* ones (Castellani and Zanfei 2006; Siddharthan 2009).

Despite the recent trend in the decentralization of R&D activities, a large number of empirical studies relating to both the developed as well as developing countries, reveal that FMEs are not more R&D intensive than DEs. In most of the OECD countries FMEs are characterized by lower R&D intensities as compared to the DEs (OECD 2005). In a study of five small European countries (Austria, Denmark, Finland, Norway and Sweden), Dachs et al. (2008) find no difference in R&D intensity of FMEs and DEs. In the case of Indian manufacturing sector, overwhelming evidences suggest that the R&D intensity of FMEs is not more than that of DEs (viz. Kumar and Saqib 1996; Ray and Bhaduri 2001; Pradhan 2002; Kumar and Aggarwal 2005; Ray and Rahman 2006; Kathuria 2008; Rasiah and Kumar 2008). Interestingly, Sunil and Sharma (2013) in a recent study find FMEs to be more R&D intensive than DEs in the Indian medium and high technology industries.

3.1.3 Intensity to Import Intermediate Goods (IMIG)

The use of superior raw materials and capital equipment ensures better quality of products leading to barriers to entry (or mobility) through differentiation advantage. The quality factor is more important in the context of machinery industry, since the efficiencies of the user industries largely depend on the quality, reliability, durability, precision and overall efficiency of machinery and equipment employed by the firms. The following major explanations for the higher import orientation of FMEs over DEs are offered in the literature. First of all, FMEs normally perceive the reliability and quality of supply in the host developing country to be inferior (Rugman 1981; Hennart 1986). Secondly, even if cost, quality and reliability of supplies are the same, FMEs may prefer to obtain inputs from their parents or parents network so that the MNE systems could capture suppliers' profit and utilize economies of scale and scope in production and distribution. Third, continuing to import intermediate inputs provides opportunities for transfer pricing which may be lost with local sourcing (Jenkins 1990).

Contrary to FMEs, DEs may prefer to procure the inputs from the local producers. First of all, they may not be well equipped to bear or tackle the uncertainty of exchange rate fluctuations and hassles of importing from the international market about which they obviously have less information than the FMEs. Secondly, DEs normally operate on the lower end of the industry that may not require such inputs for which they have to depend heavily on import. The majority of the earlier studies in developing countries reveal that the FMEs are more import intensive than DEs (Jenkins 1990; Kumar and Siddharthan 1997). The latest studies on Indian manufacturing sector and the literature survey therein (Ray and Rahman 2006) report that FMEs are more import intensive than DEs.

3.1.4 Intensity to Import Disembodied Technology (IMDT)

FMEs tend to spend more on import of disembodied technology than DEs for the following reasons: (a) FMEs have better information about and ability to scan and search frontier technology that makes easier for them to import the same; (b) FMEs often buy disembodied technology from their MNE system at transfer prices which enables the latter to boost its global profit and maintain secrecy; (c) MNE system may prevent a FME to develop technology at its host location as the local R&D involves extra expenditure and risk of technology spillovers to DEs. The empirical literature on transfer of technology in developing countries suggests that FMEs spend more on import of disembodied technology than DEs (Ray and Rahman 2006; Kumari 2007).

3.1.5 Technical Efficiency (TE)

Dunning (2000) and others identify two major reasons for higher productivity/efficiency performance of FMEs as compared to DEs: First, FMEs may have access to efficiency enhancing *technology*, managerial and organizational skills and expertise from their corresponding MNEs systems; Secondly, FMEs with longer period of presence in the host country may be able to identify, evaluate and harness information, technology and skills present therein and combine these with their internal *capabilities* for enhancing their efficiency in production. In this study, we capture the efficiency by a measure of technical efficiency which for a given firm (in a given year) is defined as the ratio of its mean output (conditional on its level of factor inputs and firm effects) to the corresponding mean output given by the stochastic frontier production function (Battese and Coelli 1992). This measure of technical efficiency by design has values between zero and one.

Several studies in recent years for the developing countries report FMEs to be more productive/efficient than DEs (e.g. Takii 2004; Takii and Ramstetter 2003 for Indonesia; Kokko et al. 2001 for Uruguay; Ramstetter 1999a for East Asian countries; Chuang and Lin 1999 for Taiwan; Ngoc and Ramstetter 2004 for Vietnam; Keshari 2013 for Indian machinery industry; Kathuria 2001; Ray 2004; Goldar et al. 2004; Sasidharan and Ramnathan (2007) for Indian manufacturing sector). On the contrary, some studies (e.g. Ito 2002; Ramstetter 2003; Oguchi et al. 2002 for Malaysia; Konings 2001 for Bulgaria and Rumania) suggest that FME are not more productive than DEs.

3.2 Others

3.2.1 Product Differentiation

Advertising and marketing tactics are considered as the two major elements of non-price strategies followed by MNEs for differentiating their products and competing with their rivals. Thus being part of MNE system, FMEs are also expected to follow more intensive advertising and marketing strategy to promote sales of their products than what is followed by DEs. Against this logic, one may also expect FMEs to be pursuing less intensive advertising and marketing strategies than those adopted by DEs in the IMI for the following reasons: (i) In the international as well as Indian market, brand equity of products sold by FMEs and corporate image of MNE system may have already been established and thereby MNE system to which FMEs belong may be well known as a reputed supplier of producer goods. Therefore, it may not be necessary for FMEs to spend substantial amount on current advertising and marketing; (ii) FMEs may be concentrated in segments of machinery industry, which may not require substantial advertising and marketing campaign for the enhancement of sales. Only a small number of empirical studies have compared the product differentiation of FMEs *vis a vis* DEs in the manufacturing sector of developing countries and findings of these studies are not conclusive (Jenkins 1990; Kumar and Siddharthan 1997).

3.2.2 Export Intensity (XI)

FMEs have the following advantages over DEs in undertaking exports (Greenaway and Kneller 2007; Kneller and Pisu 2007): First, FMEs' access to superior technology and organisational and management practices leads to higher productivity,[5] cost competitiveness, better quality and quick delivery of their products and after sale services. Secondly, production and marketing network of the MNE system itself provides an outlet for the intermediate and final products of FMEs. Thirdly, entry in third country export market requires incurring sunk cost. Since MNEs are better placed than DEs in terms of financial resources and have already incurred major part of sunk cost by virtue of multinational scope of their operation, FMEs may find it easier (than DEs) to penetrate in the international market, particularly in the markets with high barriers to entry or of highly differentiated and technologically sophisticated products. Fourthly, FMEs are better equipped to resist protectionist pressures in their home countries in such a way as to favour imports from their affiliates (Helleiner 1988).

Against the above arguments, there are the following reasons to believe that the export intensity of FMEs may not be more than that of DEs. First of all, a MNE operates with the help of its worldwide network so as to maximise the global profits but not necessarily the profits of its individual subsidiaries (Hymer 1976). Thus, a parent MNE, which has control over its FMEs, may not allow them individually to maximise exports and profits resulting from their exports, if these are expected to reduce the MNE's global profitability. This is sometimes accomplished by underpricing the exports from FMEs to MNE system. Secondly, technology transfer and financial agreements between the parents and their FMEs often include restrictive clauses controlling the export behaviour of the latter. Thirdly, if the nature of FDI is market seeking, export intensity of FMEs and DEs may not differ significantly.

---

[5]Finding in this indeed shows that FMEs are more technically efficient than DEs.

The recent studies on developing countries, which mostly use firm-level data and econometric techniques, indicate FMEs to be more export oriented than DEs. These studies include Ramstetter (1999a, b) on selected East and South East Asian Countries; Sun (2009), Du and Girma (2007) and Fung et al. (2008) for Chinese manufacturing; Jensen (2002) on Poland; Rasiah and Gachino (2005) for textiles and garments, food and beverages and metal engineering firms in Kenya; Rasiah (2004) for electronics exporting firms in Malaysia, Phillipines and Thailand; Chudnovsky and López (2004) for MERCOSUR countries; Ngoc and Ramstetter (2004) for Vietnam; Rasiah and Malakolunthu (2009) for electronics exporting firms in Malaysia; Wignaraja (2008) for a sample of clothing firms in Sri Lanka.

Kumar's (2005) literature survey on Indian studies reveals statistically insignificant difference in the export performance of FMEs and DEs during pre-reform period in majority of the cases. However, the Indian studies pertaining to post-reform period report mixed and industry-specific findings. Employing a cross-section spline regression method, Chhibber and Majumdar (2005) concludes when property rights devolves unequivocally to foreign owners (i.e. with majority ownership of equity) the Indian firms display higher export orientation. Siddharthan and Nollen (2004) report the export intensity of FMEs to be greater than those of DEs in the case of Indian information technology sector. Bhaduri and Ray's (2004) firm-level study finds no difference in export intensity of FMEs and DEs in the case of electrical/electronic industry. Using OLS method, Rasiah and Kumar (2008) find FMEs to be better than DEs in automotive parts industry. Ray and Rahman (2006), however, came to the conclusion that FMEs are less export intensive than the DEs belonging to the chemicals, electronics and transport equipment industries.

3.2.3 Financial Leverage (LEV)

We expect FMEs and DEs to differ in terms of financial leverage (LEV). In comparison to DEs, FMEs being part of MNE system are expected to have lower volatility in their earnings and increased access to international debt market, both of which, in turn, would enable FMEs to sustain a higher level of debt without increasing their default risk (Eiteman et al. 1998: pp. 583–606).

In contrast to the above, the following arguments suggest that the average LEV of FMEs to be lower than that of DEs: First of all, as per the Myers's (1984) *pecking order theory* of capital structure, if a firm is more profitable, it is more likely that it would finance its assets more from the internal sources (e.g. retained earnings which is part of networth of a firm), which is easier, readily available and more cost effective than the external sources. As FMEs are expected to be more profitable than the DEs, the former may retain lower LEV. Secondly, the financial and fiscal expertise derived from multinationalisation enables a MNE system to better manipulate taxation regulations across countries and reduce tax liabilities for its FME, implying FMEs can have higher non debt tax shield (NDTS) as compared to DEs (Singh and Hodder 2000). As the tax benefits of maintaining higher LEV are relatively less valuable for firms with higher NDTS, the FMEs (i.e. firms with higher NDTS) are expected to have lower LEV. Finally, firms with higher *agency costs* are expected to have lower debt levels (Doukas and Pantzalis 2003). FMEs' agency costs are expected to be higher relative to DEs due to higher auditing costs, language differences and varying legal and accounting systems (Burgman 1996). In sum, since the determinants of LEV may vary between FMEs and DEs, the former may maintain different LEV than the latter.

3.2.4 Firm's Size (SZ)

The size of a firm is a complex variable and may reflect the influence of several factors, including the amount of resources owned by a firm. Firm size is an indicator of managerial and financial resources available in the firm, and to the extent that excess resources are available, a firm will look for opportunities for expansion (Penrose 1959). Besides capturing amount of resources owned by a firm, the large size acts as an advantage in attracting bigger clients, gathering and processing of information, achieving economies of scale and scope in production and marketing, exerting political pressure and winning favours from the government (Mueller 1986: p. 139). As substantial resources and sunk cost are involved in establishing and operating in a foreign location, FMEs are likely to be larger than DEs. Some studies in East Asian countries have found that FMEs tend to be relatively large in comparison to DEs (Ramstetter 1999a; Takii and Ramstetter 2003).

3.2.5 Firm's Age (AGE)

IMI was initially established with the investments from Government of India through the formation of public sector enterprises (PSEs). As a result, the oldest and biggest firms in the IMI are a few PSEs [e.g. Hindustan Machine Tools (HMT), Bharat Earth Movers Ltd. (BEML), Bharat Heavy Electricals Ltd. (BHEL), and Bharat Heavy Plates & Vessels (BHPV)]. Yet, a major portion of the industry, being part of the high priority and high technology sector, has been open to foreign participation with minority equity holding of up to 40 % even before 1991 under the old industrial policy regime; and at least for 51 % foreign equity participation on automatic basis since July 1991 under the new industrial policy[6] (Kapila 2001: Chap. 19). Private including foreign participation in this industry has been increasing after the year 1991 at the cost of public sector participation. Hence, we may not find any significant difference in the average age of FMEs and DEs.

[6]The prime movers, boilers, turbines, combustion engines and steam generating plants; agricultural machinery; industrial machineries and machine tools have been the part of high priority sector.

3.2.6 Gross Profit Margin (GPM)

The reasons for higher profitability in case of FMEs compared to DEs may be the following. First of all, FMEs may enjoy higher technical efficiency/productivity. Secondly, customers of developing countries may also perceive products of MNEs as superior in terms of non-price attributes such as quality, technological sophistication, reliability, durability, just-in-time delivery and after-sales service. Therefore, they may not mind paying higher than market price to FMEs against the similar types of goods supplied by DEs. Finally, the group of FMEs enjoys greater protection from *mobility barriers*[7] against DEs and thereby may attain greater profitability on account of market power, notably in the knowledge-based industries (Kumar 1990).

Empirical evidence concerning the existence of profitability differential between DEs and FMEs is mixed but in majority of the cases FMEs outperform the DEs in terms of profit performance. Jenkins (1989) in his survey concluded that FMEs do enjoy higher profitability (than the DEs) based in the manufacturing sector of the developing countries, mainly on account of their productivity advantages and higher demand for their products. However, these studies are quite dated and use rudimentary methods of comparisons. Bellak's (2004) survey includes more recent studies which employ econometric methods for comparing the profit performance of FMEs and DEs. However, he too finds mixed results. He explains the reasons for mixed results in terms of differences in the quality of data used across the studies and rent shifting through the use of transfer pricing mechanism adopted by the MNEs.

Some studies in the context of East Asian countries (e.g. Wiwattanakantang 2001 for Thailand; Ramstetter 1999a, Ramstetter and Matsuoka 2001 for other ASEAN countries) suggest that FMEs enjoy higher profitability than DEs. Similarly, Anastassopoulos (2004) in the case of Greek food industry finds that the profitability of FMEs to be higher than that of DEs even after controlling for other determinants of profitability. In contrast, a study by Barbosa and Louri (2005), employing a quantile regression analysis suggests that foreign ownership ties in general do not make a significant difference with respect to performance of firms operating in Portugal and Greece. In the context of Indian manufacturing sector, Chhibber and Majumdar (1999) reveal significant association between foreign ownership and firms' profitability.

3.2.7 Market Concentration (MC)

Hymer (1976) stresses that the MNEs are prevalent in concentrated markets where the few firms command major share of the sales (Caves 1996: Chap. 4). In such

---

[7]Mobility barriers are defined as entry barriers, which not only impede fresh entry to the industry but also restrict inter strategic group mobility of the existing firms. Thus, firms in a particular strategic group may not only enjoy protection from new entrants to the industry but also from existing firms belonging to other strategic groups in the same industry (Kumar 1990).

markets, these firms are not price takers; and the best response of each seller is conditional upon the actions of other sellers. Lall (1978, 1979), Newfarmer (1983) and others suggest that the operations of FMEs are likely to increase the industrial concentration in the long-run and thereby they may be found mostly in the concentrated industries. In literature, the following factors are considered chiefly responsible for this phenomenon: (i) inefficient small firms may exit or merge in the face of increased competition from FMEs having competitive advantage over DEs; (ii) FMEs may use their privileged access to financial resources to outlast their rival by resorting to price and non-price warfare, and predatory practices; the distortions in market for firms considerably favour MNEs in buying out of local companies; (iii) the conducts of FMEs may have an indirect effect on concentration by stimulating defensive amalgamations among DEs and raising barriers to entry for new entrants. The TCI approach of FDI, however, seems to suggest that entry of MNEs creates more competition and breaks the existing oligopolistic structure, particularly in the developing countries (Hennart 2007). Therefore, it is more likely that FMEs are present in less concentrated and more efficient industries. Hence, it is difficult to predict whether firms in a concentrated industry or sub-industry will have more (or less) probability to observe as FMEs.

The method of construction and measurement of firm and sub-industry specific variables and hypotheses on the relative characteristics of firms between two groups of FME and DEs are explained in Table 1.

4 The Industry, Period, Data and Sample

IMI represents *manufacture of machinery and equipment n.e.c.* that is the division 28 in National Industrial Classification: All Economic Activities-2008 (NIC-2008). The division-28 comprises two types of machinery producing industries, namely, general-purpose machinery (or group 281) and special purpose machinery (or group 282) at three digit level of classification. Keeping in view the contextual nature of the impact of FDI, we select only one industry, the IMI, for this study. Selection of only one industry enabled us to reduce heterogeneity across industries arising out of differing product profiles, levels of product differentiation, industry specific policies, tax and tariff rates, levels of backward and forward integration, capital intensity, levels of technological capabilities, export orientations, etc. Focusing on only one industry also reduces heterogeneity in FDI, including the types and motives of FDI.

The major reasons for the selection of IMI *inter alia* is that there exists no firm-level study to the best of my knowledge that employs common sample of pooled data for the recent period and uses sophisticated econometric methods for simultaneous examination of several important aspects of comparative characteristics of DEs and FMEs in the IMI. Since machinery industry is categorized as the medium/high technology industry, FDI could contribute in this industry in a better way either by setting up Greenfield ventures or by offering latest technology, management and marketing expertise, international business contacts and market

Table 1 Measurement of variables and hypotheses

Variable	Definition/measurement	Hypotheses
FCD	Dependent variable FCD is a dichotomous additive dummy variable which takes the value 1 for FMEs and 0 for DEs. A firm is defined as a FME (or DE) if a foreign promoter holds at least 26 % (or less than 26 %) share in the paid-up capital of the company	
CAPI	Ratio of a firm's original (historical) cost of plant and machinery to its wage bill in FY	$CAPI_{FME} > CAPI_{DE}$ since the technology of production may have originated in the developed countries having abundance of capital and skills
RDI	Ratio of R&D expense to net sales in a FY	$RDI_{FME} < RDI_{DE}$ FMEs may not undertake R&D in host country due to centralization of R&D function in the headquarters or fear of loss of intellectual property in an alien location
IMIG	Ratio of a firm's combined expenses on import of intermediate goods including raw material, components, spare parts and capital goods to net sales	$IMIG_{FME} > IMIG_{DE}$ FMEs may import higher amount of intermediate goods due to their better quality or/and at higher prices to benefit their parents
IMDT	Ratio of a firm's expenses on payments of royalty and technical fees for the import of disembodied technology to net sales in a FY	$IMDT_{FME} > IMDT_{DE}$ FMEs being affiliates of foreign firms may import disembodied technology repetitively due to its suitability/ready availability or/and at higher prices to benefit their parents. The parent may also like to part the technology due to no fear of loss
TE	Data on firm and year specific technical efficiency (TE) is sourced from Keshari (2012). They are computed by estimating a Cobb-Douglas form of three inputs (labour, capital and raw material) stochastic frontier production function (SFPF) model by adopting Battese and Coelli's (1992) method involving the use of unbalanced panel data. Empirical model, method of construction of variables and estimates of SFPF and derivation of firm-specific TE are described in detail in Keshari (2012, 2013)	$TE_{FME} > TE_{DE}$ as former is expected to possess superior technology, management and organizational expertise
AMI	Ratio of a firm's expenditure on advertising and marketing to net sales	$AMI_{FME} > AMI_{DE}$ since product differentiation advantage created through advertising and marketing is considered as major factor in determining the competitive advantage of FMEs

(continued)

Table 1 (continued)

Variable	Definition/measurement	Hypotheses
XI	Ratio of a firm's export to net sales in a FY	$XI_{FME} > XI_{DE}$ FMEs may have higher XI because of its higher efficiency and better worldwide internal markets and external contacts
LEV	LEV is measured by the ratio of medium and long-term debts to net worth. The medium and long term debts of a company include the debt of over one year maturity. Net worth is the summation of equity capital and reserves and surplus, excluding revaluation reserves. The higher LEV of a firm (relative to other firms) means that it is financing greater proportion of its assets by debt than by owned fund (i.e. net worth)	$LEV_{FME} < LEV_{DE}$ Foreign firms are considered more prudent and have better access to equity finance
SZ	Natural logarithmic value of net sales of a firm in a year. This measure of firm size, instead of net sales, reduces degree of variability in size across firms and thereby avoids the problem of heteroskedasticity in the estimation of a regression equation	$SZ_{FME} > SZ_{DE}$ (based on empirical studies)
AGE	Age of a firm is measured by the difference between its year of presence in the sample and its year of incorporation. As every year of operation may not add significantly to the experience (or plant vintage), natural logarithm of firm's age (AGE) is taken to reduce the variability	AGE_{FME} may be equal to AGE_{DE}
GPM	Profitability of a firm is approximated by GPM which is measured by a ratio of gross profit-to-net sales of a firm in a FY	$GPM_{FME} > GPM_{DE}$ as the former may enjoy price raising capabilities based on its monopoly position and customer preferences
MC	MC is calculated as the sales weighted average of an index of a four-firm seller concentration ratio (SCR4) of each of the sub-industries of IMI in which a firm operates. The SCR4 is defined as the share of sales of four largest firms taken together in gross sales of a sub-industry of machinery industry. The procedure of calculating MC is clearly illustrated by the following example. If a firm's gross sales of Rs. 15 crore generated from sale of Rs. 10 crore worth of bearings (SCR4 = 0.90) and Rs. 5 crore worth of pumps (SCR4 = 0.30), MC applicable to the firm would be 0.70 ($10/15 \times 0.90 + 5/15 \times 0.30$)	$MC_{FME} > MC_{DE}$ FMEs may like to concentrate in more oligopolistic industries for earning higher profit

(continued)

Table 1 (continued)

Variable	Definition/measurement	Hypotheses
Sub-industry dummy variables	To control sub-industry specific influences on FCD, we use 7 sub-industry level dummy variables. For this purpose, IMI is categorized into 8 sub-industries including prime movers, engines, boilers and turbines (SI0); fluid power equipment, pumps, compressors, taps and valves (SI1); bearings, gears, gearing and driving elements (SI2); agricultural and forestry machinery (SI3); metal forming machinery and machine tools (SI4); machinery for lifting and handling goods/humans, earthmoving, mining, quarrying, construction (SI5); machinery for food, beverages, tobacco processing, textiles apparel and leather production (SI6) and other industrial machineries (SI7). A minimum 51 % of gross sales made up from a sub-industry in a particular financial year is used as the norm for this reclassification. Thereafter, we construct 7 dummy variables, SID1, …, SID7, corresponding to 7 sub-industries SI1, …, SI7. The observations on a dummy variable (say SID1) assumes the value 1 if a sample firm belongs to the corresponding sub-industry (say SI1), otherwise 0. The sub-industry SI0 is treated as the reference sub-industry, therefore, we do not use dummy variable for this sub-industry so as to avoid dummy variable trap	Not predicted

intelligence. Hence, differences in conducts and performance of FMEs and DEs may be more discernible in the IMI than in the traditional low technology industries.

The IMI constitutes about 3.76 % weight in India's index of industrial production (base 2004/05). In the market size of IMI (approximately Rs. 90,000 crore) in the year 2006/07, the share of exports constituted only about 11 %, while the share of imports was 37 %.[8] During the post-1991 reform period of August 1991 to July 2007, IMI has been relying heavily on import of disembodied technologies, but much less on FDI, for building its competitive advantage. As a result, IMI occupied the highest share of 16.6 % in the cumulative number of foreign technological collaboration agreements (7886), followed by electrical equipment (15.9 %) and chemicals (11.2 %).[9] On the other hand, the IMI's share in cumulative inflow of

[8]Refer to Keshari (2013: p. 224).

[9]Ibid.: p. 225.

FDI (Rs. 28,364 crore) of manufacturing sector constituted only 5.1 %, which compares poorly with the shares of other medium/high-tech industries (viz. electrical equipment with 32 % and transport equipment with 14 %).[10] As a consequence, during the period of study, FMEs as a group constituted only about 20 % in the aggregate sales of this industry whereas FMEs' shares are quite high in the other closely related industries, for examples, 41 % in the automobile and auto ancillaries and 42 % in the electrical machinery.[11]

The specific time period of our study covers seven financial years (FY) 2000/01 to 2006/2007. During this period, India has become one of the most attractive destinations for FDI. The period of study is important from the point of view of Indian companies adopting better accounting standards, which has made the presentations and descriptions of financial statements more detailed, transparent, accurate and uniform across the firms (Mukherjee 2008: Chap. 3). As our study uses firm-level data originally sourced from the annual reports of the companies, these developments add additional feature to our study over the studies that have used data pertaining to the period prior to the year 2000. The study has not included the period after 2006/07 as the use of longer period could lead to instability in estimated slope coefficients, particularly in view of the adverse developments in the world economy including Indian economy due to sub-prime crisis.

The major portion of the data and information is sourced from the PROWESS—an electronic database on information about the financial statements and various other aspects of Indian firms designed by the Centre for Monitoring the Indian Economy (CMIE). We also acquire data from CMIE's *Industry Market Size and Share* chiefly for constructing MC. Further, we also use some price deflators for which data is collected from various publications of the GoI. For each year of analysis, we compile relevant product/industry-wise data on *Wholesale Price Index* (base year 1993–94) from the WPI series published by the Office of Economic Advisor (OEA), GoI. Similarly, we access year-wise data on the *All India Consumer Price Index Numbers (General) for Industrial Worker* (base year 1982) from the Labour Bureau, GoI. With the help of these raw data, we design appropriate firm-level and sub-industry level variables.

We extract a list of all firms belonging to the IMI available in PROWESS database. Thereafter, we include all those firms in the sample for which data on each of the relevant variables are available for at least 2 years of the 7 financial years of the study. Further, we deleted sick companies, i.e., the companies with negative networth in a financial year, mainly with a view to remove outlier effect from the analysis. These exclusions left us with a usable sample of unbalanced panel of 177 firms with 936 observations spread over the 7 years period 2000/01 to 2006/07. Thus, the size of overall sample (as well as the size of each sub-sample of DEs and FMEs) varies from year to year during the period 2000/01 to 2006/07 of

---

[10]Ibid.: p. 225.

[11]These shares are calculated from the data obtained from PROWESS on mean of net sales of each firm for the maximum 7 years and minimum 2 years period between 2000/01 and 2006/07.

the study. Despite sample size being smaller than that of the PROWESS database, share of sample firms in respect of some aspects of corporate financial indicators (say sales turnover or net worth) of the IMI during the period of the study ranges from 66 to 90 % depending on the individual aspects of financial indicators.

Appendix Table 5 summarizes the descriptive statistics of individual variables used in the study. The descriptive statistics include mean, standard deviations (overall, between and within), minimum and maximum values of each variable. Table 5 reveals that the FCD as well as all the sub-industry specific dummy variables have no within group variation in their respective data. To know the severity of multicolinearity problem associated with the sample, we obtain variance inflation factor (VIF).[12] As a rule of thumb, if the VIF of a variable exceeds 10, that variable is deemed highly collinear (Gujarati 2004: p. 362). In terms of this rule of thumb, the values of VIF presented in Appendix Table 6 do not reveal any serious multicolinearity problem.

5 Statistical and Econometric Methods

Researchers mainly adopt three methods for comparing the characteristics of two groups of firms in an industry: a rudimentary univariate mean value method, LDA and estimation of probabilistic models (viz. logit/probit). As a preliminary exercise, we first conduct univariate analysis. Thereafter, we undertake LDA and estimate probabilistic logit and probit models. All the three methods are explained in the following sub-sections.

5.1 Univariate Method of Analysis

To compare each aspect of characteristics of two groups of firms in a univariate framework, we conduct Welch's t-test using two-samples having possibly unequal variances. To conduct this test we first of all need to calculate mean and standard deviation of individual variables for each sub-sample of FMEs and DEs. Thereafter, we are to obtain t-statistics with the help of STATA software that utilises the following formula:

[12]VIF shows the speed with which variances and covariance increase and can be defined as VIF = $1/(1 - r_{23}^2)$, where r_{23}^2 is the coefficient of correlation between X_2 and X_3. It is called variance inflating factor because it shows how the variance of an estimator is inflated by the presence of multicolinearity. If there is no colinearity between X_2 and X_3 VIF will be 1. When r_{23}^2 approaches 1, VIF approaches infinite (Gujarati 2004: Chap. 10).

$$t = \frac{\overline{X}_1 - \overline{X}_2}{s} \quad \text{where } s = \sqrt{\left[\frac{s_1^2}{n1} + \frac{s_2^2}{n2}\right]}$$

where $\overline{X}_1$ and $\overline{X}_2$ are the sample means of the FMEs and DEs respectively; s_1^2 and s_2^2 are the sample variances of the FMEs and DEs; n1 and n2 are number of observations in each group. The degrees of freedom (v) associated with variance estimates are approximated using the Welch-Satterthwaite equation. Once t and v are computed, these statistics are used with t-distribution to test the null hypotheses (H_o) that the difference in mean of a characteristic between two groups, FMEs and DEs, is zero (using a two-tailed test) against the alternative hypothesis (H_a) that the groups have different means. In other words:

$$H_0 : \overline{X}_1 - \overline{X}_2 = 0 \quad \text{against} \quad H_a : \overline{X}_1 - \overline{X}_2 \neq 0$$

A two-tail tests yields *p*-value that may (or may not) provide evidence sufficient to reject null hypothesis.[13]

5.2 *The Empirical Models of Multivariate Analysis*

The univariate mean value method compares one characteristic at a time while LDA and logit and probit models compare a firm-specific variable by controlling other firm and the sub-industry level influences. We therefore need to utilise either a multivariate framework like LDA or dichotomous probabilistic models employing multiple regression framework as both the frameworks enable us to compare a characteristic while controlling host of other characteristics. The empirical equations corresponding to the LDA, logit and probit models are presented below:

Linear Discriminant Function

$$\begin{aligned} Z = {} & b_0 + b_1 CAPI_{it} + b_2 RDI_{it} + b_3 IMIG_{it} + b_4 IMDT_{it} + b_5 TE_{it} \\ & + b_6 AMI_{it} + b_7 XI_{it} + b_8 LEV_{it} + b_9 SZ_{it} + b_{10} AGE_{it} + b_{11} GPM_{it} \\ & + b_{12} MC_{it} + b_{13} SID1_{it} + \cdots + b_{19} SID7_{it} \end{aligned} \tag{1}$$

Logit regression

$$P = E\ (FCD_{it} = 1|X) = 1/\left[1 + \exp^{-Z}\right] \tag{2}$$

[13]We prefer to use two-tail test because of the possibility that the mean of a variable capturing a characteristic for FMEs may be less or more than that of DEs.

where

$$Z = b_0 + b_1CAPI_{it} + b_2RDI_{it} + b_3IMIG_{it} + b_4IMDT_{it} + b_5TE_{it} + b_6AMI_{it} + b_7XI_{it} + b_8LEV_{it} + b_9SZ_{it} + b_{10}AGE_{it} + b_{11}GPM_{it} + b_{12}MC_{it} + b_{13}SID1_{it} + \cdots + b_{19}SID7_{it} + v_{it}$$

Probit regression

$$P = E(FCD_{it} = 1|X) = 1 - f\left[-(b_0 + b_1CAPI_{it} + b_2RDI_{it} + b_3IMIG_{it} + b_4IMDT_{it} + b_5TE_{it} + b_6AMI_{it} + b_7XI_{it} + b_8LEV_{it} + b_9SZ_{it} + b_{10}AGE_{it} + b_{11}GPM_{it} + b_{12}MC_{it} + b_{13}SID1_{it} + \cdots + b_{19}SID7_{it} + v_{it})\right] \tag{3}$$

Equation (1) is estimated for conducting LDA which identifies the discriminating characteristics of two groups (say FMEs and DEs) of firms based on certain criteria. LDA is based on two restrictive assumptions, notably, the variables in each group follow a multivariate normal distribution and covariance matrices of independent variables for each group are equal. In comparison to the probabilistic methods, LDA acts as the better method when both these assumptions are satisfied (Pohar et al. 2004). These assumptions get easily violated when we use some dummy variables as the independent variables in a model. As we use a number of sub-industry and financial year specific dummy variables in our models, we need to conduct LDA with these dummy variables and perform Box's M test, as suggested by Pohar et al. (2004), to examine the suitability of LDA *vis a vis* probabilistic models.

Equations 2 and 3 represent logit and probit models respectively in which P denotes conditional expectation of FCD_{it} given X (a vector of explanatory variables) or conditional probability that a firm will appear as FME given X. The probabilistic regression models, both logit as well as probit, relate a qualitative dependent (usually dichotomous) variable to a set of continuous and/or categorical independent variables. Probit model uses a normal cumulative distribution function (CDF), whereas the logit model employs logistic CDF, to model a relationship between a dichotomous dependent variable and the independent variables. In case of this study, both models estimate the probability of observing a firm in the group of FMEs (or DEs). Thus the positive sign of the estimated coefficient of an independent variable in these models will denote that the variable increases the probability of a firm to appear as FME.

6 Results and Discussions

6.1 Univariate Analysis

Table 2 summarizes the results on mean, standard deviation and tests of equality of group means of FMEs and DEs with respect to 11 firm specific variables representing various firm-level characteristics.

Table 2 Comparing characteristics of FMEs and DEs-univariate method (tests of equality of group means)

Variable	Domestic enterprises			Foreign multinational enterprises			Tests of equality of group means	
	Obs.	Mean	Std. Dev.	Obs.	Mean	Std. Dev.	Welch's d. o. f.	T-stat
TE	675	0.6976	0.0777	261	0.7405	0.0835	445.23	7.176*
GPM	675	0.1800	0.1187	261	0.2175	0.1094	511.39	4.600*
SZ	675	3.1821	1.6779	261	4.0633	1.2766	619.45	8.630*
AGE	675	3.1911	0.7251	261	3.2028	0.7431	463.90	0.218
CAPI	675	4.7699	5.5087	261	4.5967	3.5243	713.20	−0.569
AMI	675	0.0331	0.0347	261	0.0254	0.0287	568.06	−3.455*
IMDT	675	0.0016	0.0052	261	0.0070	0.0104	312.36	8.070*
RDI	675	0.0032	0.0058	261	0.0043	0.0065	427.06	2.376**
LEV	675	0.3655	0.2498	261	0.2516	0.2415	489.15	−6.409*
XI	675	0.1131	0.1744	261	0.1548	0.1683	489.91	3.369*
IMIG	675	0.0705	0.0873	261	0.1513	0.1159	380.61	10.197*

Note * and ** denote significance levels at 1 and 5 % respectively

T-statistics in respect of each variable is obtained by applying the formula explained in Sect. 5.1 above. Thereafter, we test the null hypothesis that the difference in mean value of each variable between the two group of FMEs and DEs would be zero. The null hypothesis is rejected in the case of 9 variables. As compared to DEs, FMEs have greater RDI, IMIG and IMDT. As the R&D activity and use of imported technology require higher level of skill, we may assume that skill intensity of FMEs may also be greater than that of DEs. These results probably suggest that FMEs do have firm-specific ownership advantage over DEs in terms of technology. In relation to DEs, FMEs on an average spend less portion of their revenue on advertising and marketing. In other words, DEs spend more towards creation of product differentiation advantage. In comparison to DEs, FMEs are also bigger in terms of their size of operation. Results on relative AGE and CAPI indicate that FMEs and DEs do not significantly differ in terms of years of operations and choice of technique. The results also indicate that FMEs, as compared to DEs, on an average achieve greater TE, GPM and XI. As compared to DEs, FMEs are also found less financially leveraged, implying that the latter finance their operations more from owned fund than from the borrowed money. As the univariate analysis compares individual characteristic independently from the others, it is imperative to build upon the findings of univariate analysis and combine several characteristics in a meaningful predictive model.

6.2 *Linear Discriminants Analysis*

Table 3 presents the results of LDA following the Mahalanobis Distance (or D square) procedure in 4 panels. Panel A of Table 3 shows that model is significant.

Table 3 Results of LDA based on stepwise procedure

Panel A: canonical distance function				
Eigenvalue	Canonical correlation	Wilks' lambda	χ^2 (19)	Prob > χ^2
0.394	0.53	0.72	308.67*	0.00

Panel B: test of equality of group covariance matrices using Box's M		
FCD	Rank	Log determinant
0	9	−37.03
1	9	−36.57
Pooled within-groups	9	−36.39
Test results (tests null hypothesis of equal covariance matrices)		
Box's M	Approximate F (45,864,988.6)	Prob > F
471.81	10.34	0.00

Panel C: mahalanobis D squired stepwise LDA					
	Variable entered	Mahalanobis D squired statistics between FMEs and DEs	Exact F		
			Statistic	df1	df2
1	IMIG	0.71	132.87*	1	934
2	IMDT	1.24	116.63*	2	933
3	LEV	1.43	89.43*	3	932
4	SZ	1.57	73.49*	4	931
5	TE	1.66	62.19*	5	930
6	AMI	1.80	56.28*	6	929
7	XI	1.92	44.81*	8	927

Panel D: discriminant functions of FMEs and DEs								
Category	TE	SZ	XI	IMIG	AMI	IMDT	LEV	Constant
DEs	127.526	1.013	5.997	−10.357	−42.088	32.387	15.655	−48.517
FMEs	133.223	1.196	7.418	−2.438	−53.502	135.385	14.509	−55.05

Notes (a) At each step, the variable that maximizes the Mahalanobis distance between the two closest groups is entered; (b) Maximum number of steps is 38; (c) Minimum partial F to enter is 3.84; (d) Maximum partial F to remove is 2.71; (e) F level, tolerance, or VIN is insufficient for further computation; (f) * denotes significant level at 1 %

Based on Box's M test, Panel B however shows that the assumption of equality in group covariance matrices is strongly rejected. Hence, the basic assumption underlying LDA gets violated in case of our study and the use of probabilistic models is considered more appropriate. Focusing on the results incorporated in Panel C of Table 3, we find that 7 firm-specific variables: TE, SZ, XI, IMIG, AMI, IMDT, LEV ultimately turn out to be significant discriminator between FMEs and DEs in the stepwise procedure.

Panel D reports the values of the estimated coefficients associated with each of these variables in the discriminant functions of FMEs and DEs. We find that FMEs, as compared to DEs, have higher TE, XI, IMIG and IMDT. FMEs are also larger than DEs. However, FMEs have less LEV and AMI. It is to be noticed that LDA does not find GPM and RDI to be a significant discriminator between FMEs and

DEs. On the other hand, the univariate analysis has found GPM as well as RDI of FMEs to be greater than GPM and RDI of DEs. However, both the univariate analysis and LDA show that the AGE and CAPI are not significant discriminators between FMEs and DEs.

6.3 Probit and Logit Results

We estimate the probit and logit models, represented by the Eqs. (2) and (3) in Sect. 5.2, by maximum likelihood technique with the help of STATA software. We also obtain heteroskedasticity-corrected standard errors by following White-Huber method with the help of robust option available in the software. Table 4 presents the results obtained from these estimations. We may note at the outset that the estimated logit and probit models offer similar results. The values of pseudo R^2 show that both the logit and probit models achieve same value of 0.26, implying one cannot differentiate between these models on the basis of overall goodness of fit. The values of Wald chi^2 and corresponding *p*-value of zero suggests that the each model (probit as well as logit) as a whole is statistically significant, as compared to the model with no regressors. Thus, there is a little to choose between probit and logit approaches.

The results on the explanatory variables show that the coefficients of 5 firm-specific variables CAPI, RDI, AGE, GPM, MC and all the sub-industry-specific dummy variables are statistically insignificant. On the other hand, the coefficients of IMIG, IMDT, TE, SZ, XI are positive and significant and coefficient of LEV and AMI are negative and significant in both the models. Comparing the results of LDA against the results of probabilistic models, we find that: (a) GPM and RDI are not discriminating factors between FMEs and DEs in LDA. Similarly, GPM and RDI do not impact the probability of a firm to appear as FME in the presence of other variables in both the probabilistic models; (b) AGE and CAPI do not differ significantly between FMEs and DEs in the LDA as well as in the probabilistic models; (c) the signs of the statistical significant coefficients of TE, SZ, XI, IMIG, IMDT, AMI, LEV are identical in both types of analysis. In sum, the results obtained from LDA and the estimation of probabilistic models are the same.

The estimation results of probit model on the factors that influence the probability of being a firm in foreign ownership also gives marginal effects (Table 4). The marginal effects are calculated for discrete change of dummy variable from 0 to 1 at the sample means and measured in terms of absolute value of a coefficient. Among the statistically significant explanatory variables, the IMDT has the greatest effects followed by AMI, IMIG, TE, XI, LEV and SZ in descending order. IMDT with the highest positive marginal effect indicates that the likelihood of being FME is the greatest for a firm that makes higher payment (as a ratio of its sales) for import of foreign disembodied technology. This result is in line with the findings of two Indian studies (e.g. Kumari 2007; Ray and Rahman 2006). This could be a reflection of import of better technology through intra-firm transactions or an indication of over payment for technology for appeasing parents. Inflation of

Table 4 Logit and probit models: estimation results

Expl. Variable	Logit model			Probit model-1			Probit model-2		
	Coef.	Het. corr. Std. Err.	z-stat	Coef.	Het. corr. Std. Err.	z-stat	dF/dx	Het. corr. Std. Err.	z-stat
TE	6.185	1.324	4.67*	3.728	0.761	4.90*	1.120	0.228	4.90*
GPM	−0.251	1.042	−0.24	−0.205	0.579	−0.35	−0.062	0.174	−0.35
SZ	0.194	0.076	2.55*	0.122	0.042	2.90*	0.037	0.013	2.90*
AGE	−0.004	0.004	−0.99	−0.003	0.002	−1.35	−0.001	0.001	−1.35
CAPI	−0.030	0.019	−1.64	−0.016	0.011	−1.51	−0.005	0.003	−1.51
AMI	−10.645	2.811	−3.79*	−6.383	1.586	−4.02*	−1.918	0.481	−4.02*
IMDT	81.725	16.623	4.92*	44.564	9.347	4.77*	13.394	2.871	4.77*
RDI	−11.963	13.775	−0.87	−7.511	8.229	−0.91	−2.257	2.476	−0.91
LEV	−1.321	0.450	−2.93*	−0.669	0.240	−2.79*	−0.201	0.071	−2.79*
XI	1.140	0.555	2.05**	0.757	0.301	2.51*	0.228	0.091	2.51*
IMIG	6.525	1.055	6.19*	3.675	0.594	6.19*	1.104	0.181	6.19*
MC	−0.727	0.875	−0.83	−0.518	0.436	−1.19	−0.156	0.132	−1.19
SID1	0.295	0.560	0.53	0.103	0.279	0.37	0.032	0.088	0.37
SID2	0.286	0.570	0.50	0.127	0.287	0.44	0.040	0.093	0.44
SID3	−0.551	0.648	−0.85	−0.299	0.316	−0.95	−0.081	0.075	−0.95
SID4	0.042	0.468	0.09	−0.062	0.246	−0.25	−0.018	0.071	−0.25
SID5	−0.656	0.522	−1.26	−0.438	0.264	−1.66	−0.117	0.063	−1.66
SID6	−0.181	0.555	−0.33	−0.060	0.287	−0.21	−0.018	0.083	−0.21
SID7	−0.001	0.494	0.00	−0.034	0.247	−0.14	−0.010	0.073	−0.14
Const.	−5.811	1.146	−5.07*	−3.423	0.636	−5.38			
Number of observations			936	Number of observations			936		
Wald Chi^2 (19)			193.88	LR Chi^2 (19)			228.39		
Prob > chi^2			0.00	Prob > chi^2			0.00		
Pseudo R^2			0.26	Pseudo R^2			0.26		
Log likelihood			−407.56	Log likelihood			−408.63		

Note *, ** denote level of significance at 1 and 5 % respectively
dF/dX is for discreet change of dummy variable from 0 to 1

payment on royalty and technical fee by FMEs has been used as a means for reducing local taxes in the host country and transferring earned profit out of the host country (Bellak 2004). In addition, the higher intensity of payment for import of disembodied technology by FMEs, coupled with no difference in R&D intensity of FMEs and DEs, imply that FMEs not only rely more on foreign technological know-how but also do not make major attempts to adapt or absorb the imported technology. To draw a firm conclusion on these issues, we need further investigation, which is beyond the scope of this paper.

The second most important factor explaining probability of a firm to be in foreign ownership is AMI. The significantly negative coefficients of AMI observed in the estimated probit and logit models show that the FMEs spend less for creating product differentiation advantage than DEs. FMEs operating in India, being part of the established international network, gain from the spillovers of the worldwide advertisements and global marketing efforts of their corresponding MNE system. As a result, they do not require to spend much for boosting their corporate image and brand equity of their products in the Indian market. Another reason could be that the threat from the entry of large number of MNEs after liberalisation from 1991 has forced oligopolistic DEs to spend heavily on advertising and marketing for the protection of their market share (Ray and Rahman 2006).

The third factor is IMIG. We may interpret the result on this aspect of firm characteristics as follows: (a) FMEs prefer to use larger amount of imported intermediate goods as these are either unavailable in the domestic market or goods available in the domestic market are inferior in their perception; (b) FMEs' may be procuring intermediate inputs from their MNE systems at transfer prices, which are higher than the market prices, for boosting the global profit of the MNE system. Our finding on IMIG is in line with the latest findings in the Indian studies (e.g. Ray and Rahman 2006).

Our finding on TE is consistent with the prediction of internalization (or transaction cost) approach of FDI and findings of several empirical studies that FMEs are more productive/efficient than the DEs (viz. Keshari 2013 for IMI; Ray 2004 for a sample of Indian manufacturing sector firms; Goldar et al. 2004 for Indian engineering industry). Combining this result with the insignificance of the coefficients of GPM and MC in the estimated logit/probit models (as well as LDA), we can conclude that FMEs are more efficient than DEs due to the efficiency in utilisation of inputs of production but FMEs do not enjoy monopoly profits.

Significant and positive coefficient of XI suggests that FMEs are not only selling in the Indian market, but also have gained competitive advantage over DEs on the export front by using the India's locational advantages as well as the advantages of being part of the MNE system. This finding is consistent with the findings of the larger set of latest Indian studies (e.g. Siddharthan and Nollen 2004; Chhibber and Majumdar 2005; Rasiah and Kumar 2008). However, our study contradicts the findings of Ray and Rahman (2006) in this respect.

FMEs are also found less financially leveraged than DEs, indicating that the FMEs use greater amount of internal funds for financing their operations. Results of our study is in line with the finding of the majority of empirical studies which report FMEs to be maintaining lower financial leverage than the DEs in the context of the developed countries (e.g. Akhtar and Barry 2009 for Japan; Chkir and Cosset 2001; Doukas and Pantzalis 2003 for USA).

Size of the firm, generally reflecting the firm's ownership of financial and non-financial resources, has positive influence on the firms' probability to appear in the group of FMEs. The reason for this could be that doing business in a foreign location also require holding of higher amount of financial and non-financial resources so as to overcome the *liability of foreignness* (Zaheer 1995). Our finding on firm size is similar to that of a study on the Indonesian manufacturing sector (Takii and Ramstetter 2005).

The coefficients of MC turn out to be insignificant in the estimated probit as well as logit model. This indicates that the probability of a firm's appearance in the group of FMEs is not dependent on the market concentration. Similarly, the coefficients of none of the sub-industry specific dummy variables are found statistically significant either in the estimated probit or logit model. These results hint that the FMEs do not show any preference for locating in one or other sub-industries of IMI. This might have happened because the sub-industries of the IMI may not be differing sufficiently in terms of overall index of characteristics so as to warrant the special attention of MNEs.

7 Conclusions and Recommendations

Our study finds probabilistic models (probit and logit) to be most suitable for the examination of technological and other characteristics of the FMEs and DEs in the IMI. Besides, in view of the similar findings from the LDA and probabilistic models, we conclude that our empirical analysis supports the proposition that the FMEs and DEs differ in terms of several aspects of technological and other characteristics in IMI. As compared to DEs, FMEs spend more on import of disembodied technology. This could be reflection of import of better technology through intra-firm transactions (or an indication of over payment for technology for appeasing parents). FMEs also spend higher amounts on import of intermediate goods including capital goods, raw material, components and spare parts. This suggests that they have fewer linkages with domestic suppliers of intermediate goods, probability because the intermediate goods used by FMEs are unavailable domestically or the domestic firms are unable to supply quality/suitable products. FMEs, however, do not spend higher amounts on R&D (as compared to DEs) which support the hypothesis on centralization of R&D in the headquarters of FMEs. In the post-WTO scenario, the regulations such as TRIMs are neither possible nor desirable for forcing FMEs to use domestic resources or undertake R&D within FMEs. Thus, the domestic suppliers of intermediate goods need to improve the quality of their products and services for encouraging FMEs to purchase their products. To encourage FMEs to spend more on R&D, the GoI needs to take steps to improve R&D infrastructure, regulatory and legal framework and implementation of IPR regime in the country so that the MNEs find India attractive enough to locate their core R&D functions.

Combining India's comparative advantage with their resource advantage and higher efficiency in production, FMEs also realize higher export intensity as compared to DEs in IMI. FMEs are able to perform better than DEs in terms of technical efficiency (but not in terms of GPM). Probably, the internal and external competition introduced in the Indian manufacturing sector (through liberal industrial, trade, FDI and associated policies followed over the years) have helped FMEs to maintain higher level of efficiency with the help of their superior *resources and capabilities* but the same has also prevented FMEs to exercise monopoly power in the market. Hence, we recommend that the outward orientations in economy with liberal and transparent FDI policy in IMI needs to be continued. This would not only contribute towards the indigenous production of additional and better machinery products with the help of FMEs but also check the influx of import and improve the efficiency of IMI. In sum, the implementation of current FDI policy will strengthen the "make in India" efforts of the Government of India.

Appendix

See Tables 5 and 6.

Table 5 Descriptive statistics of variables for full sample, 2000/01 to 2006/07

Variable		Mean	Std. Dev.	Min	Max
FCD	Overall	0.2788	0.4487	0.0000	1.0000
	Between		0.4301	0.0000	1.0000
	Within		0.0000	0.2788	0.2788
TE	Overall	0.7096	0.0816	0.5377	0.9934
	Between		0.0838	0.5447	0.9932
	Within		0.0028	0.7025	0.7156
GPM	Overall	0.1904	0.1173	−0.4871	0.7081
	Between		0.0979	−0.1754	0.4736
	Within		0.0683	−0.2759	0.6389
SZ	Overall	3.4278	1.6245	−0.1372	8.8828
	Between		1.5575	0.2772	8.5254
	Within		0.2773	2.1015	4.9944
AGE	Overall	3.1944	0.7298	0.0000	4.6250
	Between		0.7373	0.8959	4.6000
	Within		0.1266	2.0978	3.8896
CAPI	Overall	4.7216	5.0334	0.2844	50.0000
	Between		5.0590	0.3259	39.5469
	Within		1.2665	−4.5606	15.1747
AMI	Overall	0.0309	0.0333	0.0000	0.2506
	Between		0.0314	0.0000	0.2197
	Within		0.0127	−0.0548	0.1597
IMDT	Overall	0.0031	0.0074	0.0000	0.0743
	Between		0.0060	0.0000	0.0372
	Within		0.0040	−0.0215	0.0547
RDI	Overall	0.0035	0.0060	0.0000	0.0398
	Between		0.0053	0.0000	0.0284
	Within		0.0027	−0.0093	0.0260

(continued)

Table 5 (continued)

Variable		Mean	Std. Dev.	Min	Max
LEV	Overall	0.3338	0.2526	0.0000	0.9863
	Between		0.2432	0.0000	0.9577
	Within		0.1070	−0.1947	0.7288
XI	Overall	0.1247	0.1736	0.0000	0.9922
	Between		0.1523	0.0000	0.7551
	Within		0.0886	−0.3857	0.6732
IMIG	Overall	0.0930	0.1027	0.0000	0.5823
	Between		0.0918	0.0000	0.4633
	Within		0.0455	−0.1904	0.4421
MC	Overall	0.4038	0.1596	0.1256	0.8955
	Between		0.1523	0.1580	0.7762
	Within		0.0568	−0.0171	0.6845

Table 6 Indicator of multicolinearity

Variable	Variance inflation factor (VIF)
SID7	4.02
SID5	3.86
SID1	3.31
SID2	2.71
SID4	2.59
SID6	2.24
SID3	1.98
SZ	1.81
MC	1.66
TE	1.63
GPM	1.60
IMIG	1.40
RDI	1.34
CAPI	1.29
AGE	1.21
LEV	1.15
AMI	1.15
IMDT	1.11
XI	1.09
Mean	1.96

References

Akhtar, S., & Barry, R. O. (2009). Determinants of capital structure for Japanese multinational and domestic corporations. *International Review of Finance, 9*(1–2), 1–26.

Anastassopoulos, G. (2004). Profitability differences between MNE subsidiaries and domestic firms: The case of the food industry in Greece. *Agribusiness, 20*(1), 45–60.

Barbosa, N., & Louri, H. (2005). Corporate performance: Does ownership matter? A comparison of foreign- and domestic-owned firms in Greece and Portugal. *Review of Industrial Organization, 27*(1), 73–102.

Battese, G. E., & Coelli, T. J. (1992). Frontier production functions, technical efficiency and panel data: With application to paddy farmers in India. *The Journal of Productivity Analysis, 3*, 153–169.

Bellak, C. (2004). How domestic and foreign firms differ and why does it matter? *Journal of Economic Survey, 18*(4), 483–514.

Bellak, C., & Pfaffermayr, M. (2002). Why foreign-owned firms are different: A conceptual framework and empirical evidence for Austria. In R. Jungnickel (Ed.), *Foreign-owned firms–Are they different?* (pp. 13–57). Houndsmill: Palgrave.

Bhaduri, S., & Ray, A. (2004). Exporting through technological capability: Econometric evidence from India's pharmaceutical and electrical/electronics firms. *Oxford Development Studies, 32* (1), 87–100.

Burgman, T. A. (1996). An empirical examination of multinational corporate capital structure. *Journal of International Business Studies, 27*(3), 553–570.

Castellani, D., & Zanfei, A. (2006). *Multinational firms, innovation and productivity*. Cheltenham: Edward Elgar.

Caves, R. E. (1996). *Multinational enterprise and economic analysis* (2nd ed.). Cambridge: Cambridge University Press.

Chhibber, P. K., & Majumdar, S. K. (2005). Property rights and the control of strategy: Foreign ownership rules and domestic firm globalization in Indian Industry. *Law & Policy, 27*(1), 52–80.

Chhibber, P. K., & Majumdar, S. K. (1999). Foreign ownership and profitability: Property rights, control, and the performance of firms in Indian Industry. *Journal of Law and Economics, 42*, 209–238.

Chkir, I. E., & Cosset, J. C. (2001). Diversification strategy and capital structure of multinational corporations. *Journal of Multinational Financial Management, 11*, 17–37.

Chuang, Y. C., & Lin, C. E. (1999). Foreign direct investment, R&D and spillover efficiency: Evidence from Taiwan's manufacturing firms. *Journal of Development Studies, 35*(4), 117–137.

Chudnovsky, D., & López, A. (2004). Transnational corporations' strategies and foreign trade patterns in MERCOSUR countries in the 1990s. *Cambridge Journal of Economics, 28*(5), 635–652.

Dachs, B., Ebersberger, B., & Loof, H. (2008). The innovative performance of foreign-owned enterprises in small open economies. *The Journal of Technology Transfer, 34*(4), 393–406.

Doukas, J. A., & Pantzalis, C. (2003). Geographic diversification and agency costs of debt of multinational firms. *Journal of Corporate Finance, 9*, 59–92.

Du, J., & Girma, S. (2007). Finance and firm export in China. *Kyklos, 60*(1), 37–54.

Dunning, J. H. (2000). The eclectic paradigm as an envelope for economic and business theories of MNE activity. *International Business Review, 9*(1), 163–190.

Eiteman, D. K., Stonehill, A. I., & Moffett, H. M. (1998). *Multinational business finance* (8th ed.). New York, USA: Addison Wesley.

Fung, H., Gao, G. Y., Lu, J., & Mano, H. (2008). Impacts of competitive position on export propensity and intensity: An empirical study of manufacturing firms in China. *The Chinese Economy, 41*(5), 51–67.

Global Business Policy Council. (2008). *New concerns in an uncertain world: The 2007 A.T. Kearney foreign direct investment confidence index*. Chicago: A.T. Kearney Inc.
GoI. (2013). *Consolidated FDI policy*. New Delhi: Department of Industrial Policy and Promotion, Ministry of Commerce and Industry.
Goldar, B., Renganathan, V. S., & Banga, R. (2004). Ownership and efficiency in engineering firms: 1990–91 to 1999–2000. *Economic and Political Weekly, Review of Industry and Management, 39*(5), 441–447.
Greenaway, D., & Kneller, R. (2007). Firm heterogeneity Exporting and foreign direct investment. *The Economic Journal, 117*(February), F134–F161.
Gujarati, D. N. (2004). *Basic econometrics*. New Delhi: Tata McGraw-Hill Publishing Company Ltd.
Helleiner, G. K. (1988). Direct foreign investment and manufacturing for export in developing counties: A review of the issues. In S. Dell (Ed.), *Policies for development* (pp. 123–153). London: Macmillan.
Hennart, J.-F. (1986). What is internalisation? *Weltwirtschaftliches Archiv, 122*, 791–806.
Hennart, J.-F. (2007). The theoretical rationale for a multinationality-performance relationship. *Management International Review, 47*(3), 423–452.
Hymer, S. H. (1976). *The international operation of national firms: A study of direct foreign investment* (Ph.D. dissertation, MIT, 1960). Cambridge, MA: MIT Press.
Ito, K. (2002). Are foreign multinationals more efficient? Plant productivity in the thai automobile industry. Working Paper 2002–19, Kitakyushu: International Centre for the Study of East Asian Development.
Jenkins, R. (1989). Transnational corporations, competition and monopoly. *Review of Radical Political Economy, 21*(4), 12–32.
Jenkins, R. (1990). Comparing foreign subsidiaries and local firms in LDCs: Theoretical issues and empirical evidence. *The Journal of Development Studies, 26*(2), 205–228.
Jensen, C. (2002). Foreign direct investment, industrial restructuring and the upgrading of Polish exports. *Applied Economics, 34*(2), 207–217.
Jungnickel, R. (2002). *Foreign-owned firms: Are they different?*. Houndmills: Palgrave-Macmillan.
Kapila, U. (Ed.). (2001). *Indian economy since independence*. New Delhi: Academic Foundation.
Kathuria, V. (2001). Foreign firms, technology transfer and knowledge spillovers to Indian manufacturing firms—A stochastic frontier analysis. *Applied Economics, 33*(5), 625–642.
Kathuria, V. (2008). Impact of FDI inflows on R&D investment by medium and high-tech firms in India in the post reform period. *Transnational Corporations, 17*(2), 45–66.
Keshari, P. K. (2011). Comparative performance of foreign controlled and domestic firms in the Indian non-electrical machinery industry: A micro-level study. PhD thesis, Jawaharlal Nehru University, New Delhi.
Keshari, P. K. (2012). Efficiency spillovers from FDI in the Indian machinery industry: A firm level study using panel data model. In T. Tripathy et al. (Eds.), *A compendium of essays in applied econometrics* (pp. 24–55). Hyderabad: IBS-Hyderabad, IFHE University.
Keshari, P. K. (2013). Technological determinants of firm-level technical efficiency in the Indian machinery industry. *Innovation and Development, 3*(2), 223–238.
Kneller, R., & Pisu, M. (2007). Industrial linkages and export spillover from FDI. *World Economy, 30*(1), 105–134.
Kobrin, S. J. (2005). The determinants of liberalization of FDI: Policy in developing countries: A cross-sectional analysis, 1992–2001. *Transnational Corporations, 14*(1), 1–35.
Kokko, A., Zejan, M., & Tansini, R. (2001). Trade regimes and spillover effects from FDI: Evidence from Uruguay. *Weltwirtschaftliches Archiv, 137*(1).
Konings, J. (2001). The effects of foreign direct investment on domestic firms: Evidence from firm level panel data in eemerging economies. *Economics of Transition, 9*, 619–633.
Kumar, N. (1990). *Multinational enterprises in India: Industrial distribution, characteristics and performance*. London and New York: Routledge.

Kumar, N. (2005). Liberalisation, foreign direct investment flows and development: Indian experience in the 1990s (pp. 1459–1469). *Economic and Political weekly*, April 2, 2005.

Kumar, N., & Aggarwal, A. (2005). Liberalization, outward orientation and in-house R&D activity of multinational and local firms: A quantitative exploration for Indian manufacturing. *Research Policy, 34*(4), 441–460.

Kumar, N., & Saqib, M. (1996). Firm size, opportunities for adaptation and in-house R&D activity in developing countries: The case of Indian manufacturing. *Research Policy, 25*(5), 712–722.

Kumar, N., & Siddharthan, N. S. (1997). *Technology, market structure and internationalization*. New York and London: Routledge.

Kumari, A. (2007). Liberalisation, TRIPS and technological behaviour of firms in Indian pharmaceutical industry. *International Review of Business Research Papers, 3*(1), 12–22.

Lall, S. (1978). Transnationals, domestic enterprises and industrial structure in LDCs: A survey. *Oxford Economic Papers, 30*(2), 217–248.

Lall, S. (1979). Multinationals and market structure in an open developing economy: The case of Malaysia. *Weltwirtschaftliches Archiv, 115*, 325–348.

Lall, S., & Narula, R. (2004). Foreign direct investment and its role in economic development: Do we need a new agenda? *European Journal of Development Research, 16*(3), 447–464.

Meyer, K. E., Estrin, S., Bhaumik, S. K., & Peng, M. W. (2009). Institutions, resources, and entry strategies in emerging economies. *Strategic Management Journal, 30*, 61–80.

Mueller, D. C. (1986). *Profits in the long run*. Cambridge: Cambridge University Press.

Mukherjee, D. D. (2008). *Credit appraisal, risk analysis and decision making: An integrated approach to on and off balance sheet lending*. Mumbai: Snow white Publications Pvt. Ltd.

Myers, S. C. (1984). The capital structure puzzle. *Journal of Finance, 30*, 575–592.

Newfarmer. (1983). Multinationals and marketplace magic in the 1980s. In C. P. Kindleberger & D. Audretsch (Eds.), *The multinational corporation in the 1980s*. London: MIT Press.

Ngoc, P. M., & Ramstetter, E. D. (2004). foreign multinationals and local firms in Vietnam's economic transition. *Asian Economic Journal, 18*(4), 371–404.

OECD. (2005). Internationalisation of R&D: Trends, issues and implications for S&T policies: A review of literature. Presented at the Forum on the Internationalisation of R&D, Brussels.

Oguchi, N., Amdzah, N. A. M., Bakar, Z., Abidin, R. Z., & Shafii, M. (2002). Productivity of foreign and domestic firms in Malaysian manufacturing industry. *Asian Economic Journal, 16*, 215–228.

Penrose, E. (1959). *The theory of the growth of the firm*. New York: Wiley.

Pohar, M., Blas, M., & Turk, S. (2004). Comparison of logistic regression and linear discriminant analysis: A simulation study. *Metodološki Zvezki, 1*(1), 143–161.

Pradhan, J. P. (2002). Liberalization, firm size and R&D performance: A firm level study of Indian pharmaceutical industry. *Journal of Indian School of Political Economy, 14*(4), 647–666.

Ramstetter, E. D. (1999a). Comparisons of foreign multinationals and local firms in Asian manufacturing over Time. *Asian Economic Journal, 13*(2), 163–203.

Ramstetter, E. D. (1999b). Trade propensities and foreign ownership shares in Indonesian manufacturing. *Bulletin of Indonesian Economic Studies, 35*(2), 43–66.

Ramstetter, E. D. (2003). Labor productivity, wages, nationality, and foreign ownership shares in Thai manufacturing, 1996–2000. *Journal of Asian Economics, 14*, 861–884.

Ramstetter, E. D., & Matsuoka, A. (2001). Recent trends in large firms in Singapore and Thailand: Did foreign multinationals perform differently than local firms through the crisis? In M. Toida & D. Hiratsuka (Eds.), *Projections for Asian Industrializing Region* (pp. 79–180). Tokyo: Institute of Developing Economies.

Rasiah, R., & Kumar, A. (2008). Foreign ownership, technological intensities and economic performance of automotive parts firms in India. *Asia Pacific Business Review, 14*(1), 85–102.

Rasiah, R. (2004). Exports and technological capabilities: A study of foreign and local firms in the electronics industry in Malaysia, the Philippines and Thailand. *European Journal of Development Research, 16*(3), 587–623.

Rasiah, R., & Malakolunthu, A. (2009). Technological intensities and economic performance: A study of foreign and local electronics firms in Malaysia. *Asia Pacific Business Review, 15*(2), 181–297.

Rasiah, R., & Gachino, G. (2005). Are foreign firms more productive and export- and technology-intensive than local firms in Kenyan manufacturing. *Oxford Development Studies, 33*(2), 211–227.

Ray, A. S., & Bhaduri, S. (2001). R&D and technological learning in Indian industry: Econometric estimation of research production function. *Oxford Development Studies, 29*(2), 155–171.

Ray, P. K., & Rahman, S. (2006). The propensity for local innovation and inter-firm linkages in transnational corporations versus local enterprises in India. *Transnational Corporations, 15*(2), 71–98.

Ray, S. (2004). MNEs, strategic alliances and efficiency of firms: Emerging trends. *Economic and Political Weekly, Review of Industry and Management, 39*(5), 434–440.

Rugman, A. M., & Verbeke, A. (2001). Subsidiary-specific advantages in multinational enterprises. *Strategic Management Journal, 22*, 237–250.

Rugman, A. M. (1981). *Inside the multinationals: The economics of internal market*. London: Croom Helm.

Sasidharan, S., & Ramnathan, A. (2007). Foreign direct investment and spillovers: Evidence from Indian manufacturing. *International Journal of Trade and Global Markets, 1*(1), 5–22.

Siddharthan, N. S. (2009). Determinants of R&D: Lessons from literature. Working Paper, Forum for Global Knowledge Sharing and Madras School of Economics.

Siddharthan, N. S., & Nollen, S. (2004). MNE affiliation, firm size and exports revisited: A study of information technology firms in India. *The Journal of Development Studies, 40*(6), 146–168.

Singh, K., & Hodder, J. E. (2000). Multinational capital structure and financial flexibility. *Journal of International Money & Finance, 19*(6), 853–885.

Smeets, R. (2008). Collecting the pieces of the FDI knowledge spillovers puzzle. *The World Bank Research Observer, 23*(2), 107–138.

Sun, S. (2009). How does FDI affect domestic firms' exports? *Industrial Evidence. World Economy, 32*(8), 1203–1222.

Sunil, K. A., & Sharma, R. (2013). Innovation by firms in a high and medium-high technology industries: An Indian experience. In Conference Proceedings of VIII Annual International Conference of Forum for Global Knowledge Sharing on Emerging Technologies and Development (pp. 117–144).

Takii, S. (2004). Productivity differentials between local and foreign plants in Indonesian manufacturing, 1995. *World Development, 32*(11), 1957–1969.

Takii, S., & Ramstetter, E. D. (2003). Employment, production, labor productivity, and foreign multinationals in Indonesian manufacturing, 1975–2000. Working Paper No. 2003–25, International Centre for the Study of East Asian Development, Kitakyushu.

Takii, S., & Ramstetter, E. D. (2005). Multinational presence and labour productivity differentials in Indonesian manufacturing, 1975–2001". *Bulletin of Indonesian Economic Studies, 41*, 221–242.

UNCTAD. (2014). *World investment report 2014: Investing in the SDGs—An action plan*. New York and Geneva: United Nations.

Wignaraja, G. (2008). foreign ownership, technological capabilities and clothing exports in Sri Lanka. *Journal of Asian Economics, 19*, 29–39.

Wiwattanakantang, Y. (2001). Controlling shareholders and corporate value: Evidence from Thailand. *Pacific-Basin Finance Journal, 9*, 323–362.

Zaheer, S. (1995). Overcoming the liability of foreignness. *Academy of Management Journal, 38* (2), 341–363.

Does M&A Matter for R&D? Evidence from the Pharmaceutical Sector in India

Vidhisha Vyas and K. Narayanan

1 Introduction

Growing body of literature argue that mergers and acquisitions [M&A] has become a means for firms to acquire, absorb and exploit the knowledge assets of target firm. Firms with lower innovative capabilities obtain fresh and complementary technologies via acquisition of innovative firms (Barkema and Vermeulen 1998). However, acquisition can also be an attractive strategy for R&D intensive firms lacking specific knowledge (Hennart and Park 1993). While market seeking strategies are main driving force for external investments, asset seeking motives are turning out to be more crucial criteria in a firm's decision to undertake M&A (Dalton and Serpio 1999). It is widely accepted that technological expertise, market know-how, tacit knowledge and quick innovation are crucial corporate assets for facing increased competition (Cantwell and Santangelo 2002).

V. Vyas (✉)
IILM Institute of Business and Management, Gurgaon, India
e-mail: vidhisha.d.vyas@gmail.com

K. Narayanan
Department of Humanities and Social Sciences, IIT, Bombay, India
e-mail: knn.iitb@gmail.com

N.S. Siddharthan and K. Narayanan (eds.), *Technology*,
India Studies in Business and Economics,
DOI 10.1007/978-981-10-1684-4_6

The impact of M&A on firm's innovation aspects are supported by several theories. Resource based approach asserts that in the light of rising competition, M&A turns out to be an important vehicle through which firms can augment their asset base by avoiding time consuming internal processes of accumulating innovating resources (Barney 1991; Teece et al. 1997). Through acquisitions, firm specific asset with one organization are used more productively in combination with assets of another organization (Anand and Singh 1997; Capron et al. 1998). Theory of Industrial organization supports the argument that M&A provide firms an opportunity to reap benefits of economies of scale and scope via cost saving and risk spreading strategies and allow them to carry out multiple R&D projects simultaneously (Henderson and Cockburn 1996). Theory of learning and innovation suggests that firms can develop their knowledge base by investing in multiple knowledge enhancing projects however during the same period firms can grow their knowledge base by acquiring, absorbing and assimilating the external knowledge bases (Cohen and Levinthal 1989; Huber 1991). However, the theory of corporate control suggests that M&A will hamper innovation activities of firms due to agency problems (Haspeslagh and Jemison 1991), reduction in managerial commitment for R&D projects, consumption of managerial time and energy in integration process (Hitt et al. 1991, 1996) and low retention rate of key inventors (Ernst and Vitt 2000).

M&A is turning out to be an important phenomenon for corporate restructuring and facing global competition, however, little academic research has been devoted to study the impact of M&A on innovative performance in emerging economies like India. The present paper aims to explore this under investigated topic by examining the impact of M&A on inputs to the innovation process that is R&D intensity. The study also compared whether the findings of similar studies on European and American companies should continue to hold true for emerging economy like India. The methodology adopted is propensity score approach (Rosenbaum and Rubin 1983) to account for endogeneity of decision to undertake M&A to acquirer's characteristics that are correlated with post-acquisition innovation process. Propensity score enables us to remove potential endogeneity to observable firm characteristics by creating counterfactual innovation performance (i.e. innovative performance of a firm in situation of non-occurrence of M&A event).

Our analysis also focuses on characteristics of acquirers to explain the impact of M&A outcomes on firm's innovative process. The appropriate acquirer's characteristics for the analysis have been drawn from learning and innovation approach (Cohen and Levinthal 1989; Makadok 2001; Narayanan 1998) and financial economics approach (Jensen 1986). We also used deal specific characteristics for the study in the form of control variables like relatedness of acquisition and geographical location of the deals.

Diagrammatic Representation for ex-Post Impact of M&A on R&D Intensity

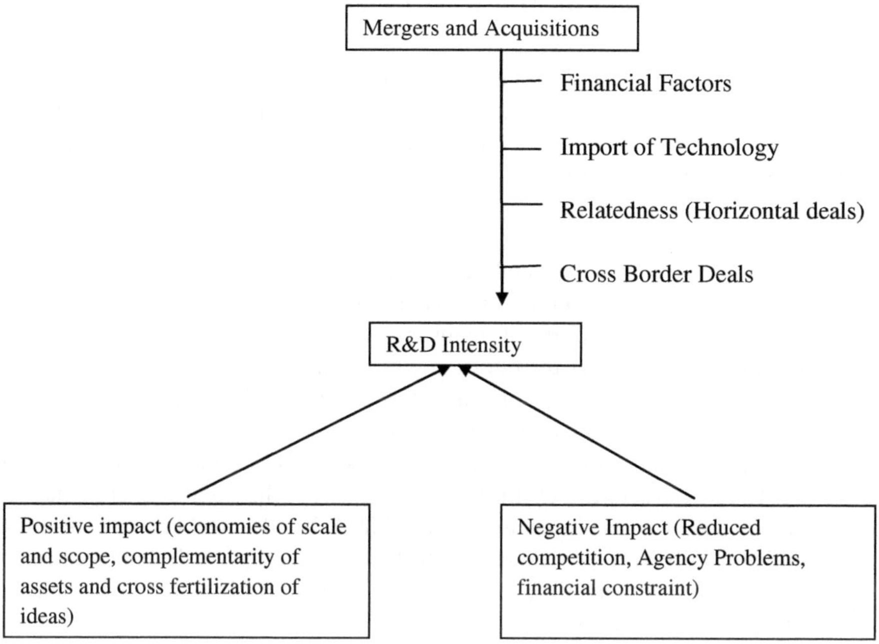

Pharmaceutical industry is chosen for the study because it has played a prominent role in the context of Indian Economy. The Indian pharmaceutical industry is ranked 3rd in the world in terms of production volume (10 % of global share) and 10th in terms of value. The industry has played a prominent role in context of Indian M&A wave accounting for some of the big M&As. Also, this industry has high R&D intensity and capacity to innovate, which is clearly the most important factor to compete among other firms.

The M&A phenomenon is under researched in terms of developing nations. This paper contributes to the existing literature in the sense that it provides useful insights on the post M&A R&D behavior of firms in Indian pharmaceutical sector.

The remaining paper proceeds as follows. Section two presents the theoretical underpinning and the review of literature pertinent to the research. Section three continues with data and variable description and methodology of the study. Section four presents the empirical analysis and discussion of results. Section five draws the concluding remarks.

2 Review of Literature

2.1 Ex-Poste Impact of M&A on R&D

Acquisition is a means of external technology sourcing which can be complementary, substitutive or both to in-house R&D. Innovation in high technology industry has become increasingly a medium to survive in highly competitive markets. As a result large firms frequently face make or buy decision dilemma, especially for quicker innovation (Wagner 2011). This establishes acquisition as front runner in technology sourcing methods. Research has proved that acquisition strengthens innovation and financial performance of the firms (Prabhu et al. 2005; Gantumur and Stephan 2007) but at the same time other set of research concluded that M&A reduces R&D efforts of acquiring firms (Hall 1990; Hitt et al. 1991). Therefore, we tried to answer the question whether acquisition affect the acquirers' innovation performance in terms of innovation inputs i.e. R&D intensity.

According to industrial organization literature, M&A are associated with economies of scale and scope of R&D and production as well as internalization of spillovers. R&D restructuring by firms in the light of technological change has led to increase in R&D costs. In such scenario M&A besides seeking new business opportunities also behave as risk spreading tactics which partially balances rising R&D costs. M&A overcome time and cost constraints of R&D activity by acquiring technological and human resources. The rising number of M&A in pharmaceutical sector is a solution to cope with rising R&D costs attached to the production of new drugs and shortening of the pipeline gap. However, in post-acquisition period firms eliminate duplication of R&D inputs leading to reduction in R&D expenditure in short term. M&A are also usually accompanied by large bureaucratic cost causing delay in decision of new R&D projects. Post M&A integration problems and diversion of manager's attention along with financial constraints leads to lower R&D investments.

On the other hand, M&A facilitates more promising and high cost R&D programs since there will be greater availability of internal finance. Size benefits also provide easy access to financial markets (Bertrand and Zuniga 2006). Increased complementarity of technological assets of parties involved in M&A help in improving R&D efficiency. The complementary knowledge acquired by M&A promotes innovation by cross fertilization of ideas between acquirer and target firm's personnel (Bertrand and Zuniga 2006). Post M&A, firms can reap benefits from economies of scale in non-R&D activities like in output production and distribution. Therefore, according to Schumpeterian hypothesis when R&D investment is spread over larger output a positive effect on R&D investments can be anticipated (Veugelers 2008).

The relationship between M&A and innovation performance has been investigated by several studies in the past. They focused their analysis on proxies of either R&D inputs or on R&D output (Danzon et al. 2007; Healy et al. 1992 Ravenscraft and Scherer 1987 on R&D input and Chakrabarti et al. 1994; Ahuja and Katila 2001; Hagedoorn and Duysters 2002) on R&D output). But there are some studies which offer insight on both R&D input as well as on R&D output simultaneously (Desyllas and Hughes 2010; Gantumur and Stephan 2007 on both). Hall (1990) examined the impact of corporate restructuring on industrial research spending and concluded a permanent decline in R&D intensity of acquiring firms. Hitt et al. (1991) analysed the acquisition effect on R&D intensity and patent intensity for a sample of 191 US firms and reported significantly negative impact on both the innovation parameters. Even in another qualitative study, Hitt et al. (1996) obtained negative impact on R&D intensity and output. Bertrand (2009) analysed acquisition of foreign firms by French firms and reported an increase in R&D spending in post-acquisition years. While Ornaghi (2009) in his study of 27 large pharmaceutical M&A concluded that mergers do not deliver expected innovative efficiency. Ahuja and Katila (2001) and Cloodt et al. (2006) reached similar conclusion asserting that M&A boosts R&D output of acquiring firms. Desyllas and Hughes (2010) report in their results negative to positive effect on R&D intensity and negative to neutral effect on R&D productivity.

2.2 *Technology Imports and Development of Absorptive Capacity*

Firms acquire technology externally either through imports (directly in the form of embodied technology in capital goods or at arm—length purchase by paying lump sum fees and royalty payments) or in the form of foreign direct investment or through M&A activities. Following acquisitions, in house R&D efforts are required to locate, adapt assimilate, and develop the acquired technology for ready use. Therefore, technology acquisitions need to be complemented with in—house R&D efforts (Narayanan 1998). Cohen and Levinthal (1989) in their seminal paper explained that in—house R&D is required not only to pursue new product and process innovation but also to assimilate and exploit externally acquired technology.

Firms with better absorptive capabilities are considered to be more judicious in carrying out M&A activity. Makadok (2001) explained two aspects of absorptive capacity. One is 'resource picking' in which firms with absorptive capacity are better positioned to screen the target and choose appropriate one and at the same time discourage them to choose inappropriate targets. Along with resource picking it is important for firms to exploit the acquired resources which are termed as 'capacity building' by Makadok (2001).

Before liberalization and many years in post liberalization era technology transfer or innovation in India were largely in the form of technology imports and these imports facilitates technological paradigm shifts in Indian economy (Narayanan 1998). Technology paradigm shift enable firms to operate on different technology frontier. The knowledge base of Indian firms is largely based on technological imports and their adaptability by performing in-house R&D. Under such circumstances, the earlier import of technology can be considered as stock of knowledge for the firms and subsequently develops their absorptive capacity.

The technology procured via imports can be insufficient but have created a substantial technological base for countries like India and therefore; M&A can provide requisite technology to the firms for developing their innovative capabilities. In the light of the above arguments it could be hypothesized that the impact of technology imports of the acquiring firms on R&D intensity can turn out to be positive.

2.3 Financial Capacity

Hall (1990) established in her study the link between leverage and reduced R&D expenditure of M&A firms. High leverage does not favor investment in R&D because high leverage entails higher financial risks. In post-acquisition period managers try to minimize volatility by avoiding investments in risky, long horizon payback R&D projects and seek stable source of profit in order to ensure steady cash flow for repaying debts (Hitt et al. 1996). The assets created by investment in R&D projects are often not re-deployable and seldom transferable. Even the human capital associated is also project specific. Firm's R&D capabilities decreases with active acquisition phenomenon. This effect can be attributed to transaction cost involved and to post acquisition integration task which absorbs managers' time and energy. Smith and Warner (1979) also assert that focus on acquisition and high level of debt possibly limits managers' discretion either because stock holders and debt providers imposed strict limitations for their funds or managers become risk-averse. Therefore, asset specificity and cash flow argument suggest that leverage taken for M&A will have negative impact on R&D intensity.

Alternatively, Myers and Maljuf (1984) argued that high level of leverage might not impact R&D intensity of acquiring firms adversely because firms first utilize internal cash flow followed by debt and finally when leverage is maximized issue fresh equity for financing acquisitions. Therefore, leverage might or not be a constraint for expenditure on R&D. It may be a case that leverage growth at the time of acquisition could restrict resource allocation to R&D projects.

In light of above argument we can hypothesize that high level of leverage or high leverage growth at the time of acquisition will adversely impact R&D intensity of acquirers.

2.4 Related Acquisitions (Horizontal M&A)

Horizontal acquisition considers that target and acquirers are operating in similar markets. It is easier to reap synergistic benefits when two firms are operating in one industry (Chakrabarti et al. 1994; Ahuja and Katila 2001). In acquisitions, greater the knowledge base concentration of the acquirer, the greater has its expertise in specific technology fields hence it can identify an appropriate related target and could effectively exploit its acquired technology and knowledge (Prabhu et al. 2005). Hagedoorn and Duysters (2002) studied relationship between M&A and technological performance of computer industry of developed economies. They suggested that related M&A has higher technological performance than unrelated M&A. Cassiman et al. (2005) hypothesized that the impact of M&A on R&D process depends upon relatedness (technological or market) of acquirer and target firms. Capron and Mitchell (1998) indicated that the efficiency of horizontal acquisitions by bilateral resource redeployment and asset divestiture might enhance firm's technical capabilities and hence stimulates their R&D efforts. This study, therefore, hypothesizes a positive relationship between horizontal M&A and R&D intensity.

2.5 Cross-Border Acquisitions

Cross-border M&A deals are one of the most important examples of industrial globalization. All the industries have witnessed increasing number of cross-border deals and pharmaceutical sector is no exception to it. Cross border deals generate technological complementarities fostering diffusion of knowledge between the dealing firms. Innovative capabilities are promoted because geographical locations creates heterogeneity between acquiring and target firm which is reflected in terms of labour and capital endowment and economic and regulatory environment of two countries (Bertrand and Zuniga 2006). On the other hand Kogut and Zander (1992) specifies that cross border deals could lead to higher integration cost. Cultural and geographical distances hamper technology transfer by making communication as well as assimilation of acquired knowledge difficult.

Gugler et al. (2003) did not find any significant impact of cross border deals on profitability of firms while Markides and Ittner (1994) found cross border deals to be welfare improving for US acquirers. Bertrand and Zuniga (2006) concluded in their study that domestic M&A has enhanced R&D investment in low-technology industries but cross border M&A has insignificant impact on R&D investment in all group of industries. Desyllas and Hughes (2010) also confirmed negative impact of cross border deals on R&D processes of the acquiring firms and attributed this negativity to higher integration and regulatory costs. In the case of Indian Pharmaceutical industry, firms may use cross-border acquisitions as a substitute for technology imports involving tacit information and therefore, acquisitions could positively enhance R&D intensity. In this paper, the types of M&A [horizontal and cross-border] are used more as control variables to account for differences between general M &A and specific ones.

3 Data, Variables and Methodology

3.1 Data Description

The study uses pooled cross-sectional data, for the period from 2000 to 2010 for pharmaceutical sector in India. The source of data for M&A deals and firm characteristics is CMIE Prowess database version 4. The number of the firms in each year is 171, with a total of 1360 observations for 8 years. To conduct empirical analysis of M&A effect on R&D intensity for three post acquisition years we used M&A deals which took place till the year 2007. For dependent variable R&D intensity and other firm characteristics like leverage and import of technology we extended data set till the year 2010. The sample firms with acquisition activity carried out 134 M&A during the period of 2000–2007 and where the firms make more than one acquisition in a given year we treat that as only one "acquirer" in that year in the present analysis.

3.2 Variables Description

3.2.1 Dependent Variables

Percentage Change in R&D intensity (R&D)—We tried to measure innovative performance using data on R&D expenditure. We constructed dependent variable in the form of R&D intensity measured by the ratio of R&D expenditure to net sales. By normalizing R&D expenditure by a proxy of firm size we make sure that our variable is not affected by change in size (due to sales) on yearly basis. The percentage change in R&D intensity is calculated from $t-1$ to $t+1$, $t+2$ and $t+3$ respectively. We also calculated change in R&D intensity from $t-1$ to the three year average R&D intensity over the period from $t+1$ to $t+3$. Several studies has used three post acquisition year window to analyze the impact of M&A on innovation performance as well as on economic performance (Ahuja and Katila 2001, Bertrand and Zuniga 2006; Ornaghi 2009; Desyllas and Hughes 2010). Post-acquisition three year analysis allows firms to integrate target in effective way up to some extent where it is anticipated that effect of M&A can be significantly visible.

3.2.2 Independent Variables

Mergers & Acquisitions (DMA)—It is a binary variable that takes value one in years when a firm makes at least one acquisition and takes value zero otherwise during the period of 2000–2007.

Import of technology—Import of technology is captured by direct import of capital goods called as embodied technology import or through arm's length by paying royalty and lump sum fees called as disembodied technology imports. We captured embodied technology intensity *(ETI)* by ratio of expenditure on imports of capital goods to net sales. Disembodied technology intensity *(DTI)* variable is computed by ratio of Lump sum, royalty, and technical fees payments in foreign currency to net sales.

Leverage (LEV)—Leverage is measured by the ratio of total borrowing of the firms to the total assets of the firms. We captured leverage growth by measuring the change between the last pre-acquisition year and first post-acquisition year. We assumed that the leverage growth *(LEVG)* for acquirer is caused by debt-financing of M&A deal.

3.2.3 Control Variables

Related Acquisitions (DHMA)—The dummy variable is introduced to discriminate between horizontal and other types of acquisitions. Horizontal acquisition considers that target and acquirers are operating in similar markets. It is easier to reap synergistic benefits when two firms are operating in one industry (Chakrabarti et al. 1994; Hagedoorn and Duysters 2002; Ahuja and Katila 2001). The dummy variable equals one where acquiring and target firm have same 3-digit NIC code and zero otherwise.

Cross-border Acquisitions (DCB)—We employed a dummy variable to discriminate between domestic and cross-border acquisitions. The dummy variable equals one for acquisitions where the target firm is incorporated in foreign country and zero otherwise.

3.3 Methodology

The primary objective of the study is to estimate the effect of M&A on R&D intensity of firms in post M&A period. For each firm i in the sample let $M\&A_i$ be a merger and acquisition indicator that equals one when the firm engages in M&A event and zero otherwise. Y_{i1} is the change in R&D intensity of M&A participating firm and Y_{i0} is the change in R&D intensity of non-M&A participating firms.

Therefore

$$Y_{i1} = M\&A_i + (1 - M\&A_i)Y_{i0}$$

Accordingly let $E\{Y_{i1}|M\&A_i = 1\}$ and $E\{Y_{i0}|M\&A_i = 0\}$ express average outcomes of innovative performance of M&A and Non M&A firms respectively. The effect we would like to examine is that of M&A on innovative performance. In other words, the difference between expected innovative performance of firms participating in M&A and what would they have experienced, if, they had not participated in M&A event.

$$\tau = E\{Y_{i1}|M\&A_{it} = 1\} - E\{Y_{i0}|M\&A_{it} = 1\}$$

This is known as expected or average treatment effect on the treated firms (Dehejia and Wahaba 2002).

Since it is not possible to find out counterfactual evidence of what would have happened had the firm not participated in M&A event, $E\{Y_{i0}|M\&A_{it} = 1\}$ is unobservable. One way to estimate the counterfactual performance is by utilizing information from firms not participating in M&A events i.e. $E\{Y_{i0}|M\&A_i = 0\}$. Therefore the effect can be estimated by difference in expected outcome between the acquiring and non-acquiring innovative performance.

$$\tau = E\{Y_{i1}|M\&A_{it} = 1\} - E\{Y_{i0}|M\&A_{it} = 0\}$$

However, observing Y_{i0} for non acquirers will result in biased estimate of acquiring firm's counterfactual performance, if acquirers and non-acquirers firms systematically differ in their firm characteristics (Hirano et al. 2003). Another cause of biasness could be observed if M&A is endogenous to certain firm characteristics and these characteristics are correlated to post acquisition performance.

To overcome this problem Rosenbaum and Rubin (1983) proposed that a propensity score analysis of similar observational characteristics can be used to create treated and control groups and subsequently post-merger performance effect can be measured using these matched groups.

The propensity score $p(M_i)$ is defined as the probability that firm i will engage in year t conditional upon observed covariates X

$$p(M_{it}) = Pr\left(M_{it} = 1|X_{i,t-1}\right)$$

If the outcomes $(Y_{i1}\&Y_{i0})$ are independent of the assignment to treated and control firms conditional on observed covariates then classifying firm observations by their propensity score balances the observed covariates X within a subclass with similar $P(M_i)$, the distribution of X is same between treatment and control groups (Rosenbaum and Rubin 1983). Further the treatment effect of M&A firms with given propensity score is estimated by difference mean outcome of treatment and control group.

$$\tau = E(Y_{it}|P(M_{it}), M_{it} = 1) - E(Y_{i0}|P(M_{it}), M_{it} = 0)$$

But in our analysis we utilized predicted probabilities i.e. propensity score weighting approach suggested by Hirano et al. (2003) and earlier used by Desyllas and Hughes (2010). They explained that weighting by the inverse of an estimated propensity score will give efficient estimate of average treatment effect which in this study is the acquisition impact on technological performance of acquirers. The present study also uses the PSM technique for weighing treated and untreated observations.

Following Desyllas and Hughes (2010) we used the following algorithm for estimating acquisition effect. At first the propensity score is estimated by running logit regression on lagged values of acquirers' economic and innovation characteristics. Use of lagged values take care of endogeneity problem associated with M&A decision and other observable firm characteristics. The dependent variable is a binary variable taking value one and zero depending upon firms' decision to participate in M&A event or not. The explanatory variables include size, leverage, disembodied technology intensity, embodied technology intensity, profit margin, dummy zero R&D, R&D intensity, growth, knowledge base size, R&D productivity measured in $t - 1$ time period. Year dummies are also included. The results of logit estimation are presented in Appendix A.

In propensity score matching a potential bias can arise from lack of overlapping or mismatching between acquirer and non-acquirer firms which occurs when some treated observations are not comparable to control observations (Heckman et al. 1997). Common support region condition helps in accounting this possible bias. We compare the maximum and minimum propensity score in acquiring and non-acquiring groups. In this process we eliminate the observations on acquirers whose propensity score is larger than maximum of non- acquirers and those of non-acquirers whose propensity score is smaller than minimum of acquirers. As a consequence eight observations fall out of our sample.[1]

To check for the adequacy of propensity score the t-test confirms that the mean of observed characteristics after matching are not systematically different for control and treated groups and mean bias has reduced to 2 % from 32.5 % (see diagram in Appendix B) Another check whether propensity score balances the observable characteristics between acquirer and non-acquirer group is performed by regressing each covariates on a dummy variable discriminating between M&A and non M&A firm observations and year dummies. The result indicated that the M&A dummy is statistically insignificant. This explains that controlling for propensity score balances the observable characteristics between acquirers and non-acquirers.

In the second stage of the analysis we carried out a weighted least squares regression by regressing percentage change in R&D intensity on a dummy variable that takes value one when firm carries out an acquisition and zero otherwise. The

[1]Ranbaxy Ltd. is the firm which does not fall under common support system and we removed all its observation for 8 years.

weight of the firm year observation for acquisition dummy equals to one is 1/p and for acquisition dummy equals to zero is 1/1 − p. Therefore, M&A firms are given more weights when they have lower propensity score while non M&A firms are given more weights when they have high propensity score.

4 Empirical Analysis

4.1 Preliminary Descriptions

The summary statistics presented in Table 1 indicates that for all the three post acquisition years M&A firms has lower change in R&D intensity in comparison to control group firms. These statistics provide preliminary suggestion that as depicted by previous literature M&A do not enhance R&D intensity of acquiring firms rather similar firms who have not participated in M&A activity are spending more on R&D projects. M&A undertaking firms has higher leverage and leverage growth in comparison to control group. The mean of disembodied technology imports is lower for M&A firms as compared to control group and mean of embodied technology imports is higher for M&A performing firms than control firms.

4.2 Weighted Least Squares Regression Result

4.2.1 The Acquisition Effect on Acquirers R&D Intensity

Results of WLS regression for estimating the impact of M&A on change in R&D intensity of acquirers are reported in Table 2. The coefficient of acquisition dummy explains the impact of M&A on R&D intensity through WLS regression after controlling for propensity to acquire. The percentage change in R&D intensity from

Table 1 Summary statistics

Variable	Mean (S.D.)	Mean (S.D.)	Mean (S.D.)
	Full sample	M&A firms	Control sample
R&D $_{t+1}$	0.24 (1.29)	0.21 (1.16)	0.25 (1.33)
R&D$_{t+2}$	0.378 (2.40)	0.29 (1.45)	0.40 (2.63)
R&D$_{t+3}$	0.39 (2.28)	0.28 (1.45)	0.43 (2.49)
R&D average	1.86 (5.72)	2.18 (3.57)	1.75 (6.26)
Leverage	0.32 (.39)	0.35 (0.55)	0.31 (0.32)
Leverage growth	-0.15 (13.7)	0.42 (3.78)	−0.34 (15.65)
DTI	0.012 (0.06)	0.0006 (0.003)	0.001 (0.006)
ETI	0.0009 (0.005)	0.017(0.075)	0.010 (0.05)
No. of observations	1360	341	1019

Table 2 Weighted least squares regressions: analysis of acquisition effect on R&D intensity

% change between t − 1 and variables	t + 1			t + 2		t + 3		Average t + 1 to t + 3
	(1)	(2)	(3)	(4)	(5)	(6)	(7)	(8)
Constant	0.142 (1.19)	0.179 (1.88)*	0.31 (1.03)	0.55 (2.75)***	0.55 (3.85)***	0.54 (3.44)***	2.03 (4.84)***	2.02 (4.60)***
DMA	0.013 (0.14)	0.265 (1.85)*	0.15 (0.35)	−0.75 (−1.79)*	−0.28 (−2.37)***	−0.60 (−2.91)***	−0.94 (−3.61)***	−1.24 (−4.29)***
DCB		−0.036 (−0.28)		5.97 (1.23)		0.40 (1.71)*		1.25 (3.59)***
DHMA		−0.226 (−1.58)		0.48 (1.36)		0.52 (2.57)***		0.57 (1.67)*
$LEVMA_{t-1}$		−0.213 (−7.16)***		0.68 (0.62)		−0.602 (−1.69)*		−0.15 (−3.90)***
$LEVGMA_{t-1}$		0.003 (0.45)		−0.004 (−0.24)		0.008 (1.61)*		0.009 (0.38)
$DTIMA_{t-1}$		−24.85 (−2.24)**		−78.63 (−1.96)**		−35.05 (−2.50)***		−67.14 (−2.28)**
$ETIMA_{t-1}$		1.58 (0.63)		−0.82 (−0.33)		2.32 (4.89)***		9.66 (1.91)**
LEV_{t-1}		0.0001 (3.37)***		0.0002 (4.35)***		0.0002 (6.39)***		0.0003 (6.00)***
$LEVG_{t-1}$		−0.0030 (−0.37)		0.010 (.53)		−0.008 (−1.57)		−0.006 (−0.26)
DTI_{t-1}		−0.76 (−0.52)		−1.54 (−0.30)		−1.87 (−0.97)		−6.05 (−1.03)
ETI_{t-1}		0.050 (0.09)		−0.34 (−0.37)		−0.60 (−1.46)		0.96 (0.52)
Time Dummy	Yes	Yes	Yes	Yes	Yes	Yes	Yes	Yes
No. of Observations	1350	1350	1350	1350	1350	1350	1350	1350
F-statistics	2.8***	42.89***	4.02***	3.70***	4.61***	12.76***	3.65***	12.43***
R^2	0.042	0.0974	0.029	0.142	0.0673	0.0983	0.017	0.0393

*** Indicates 1 %
** Indicates 5 %
* Indicates 10 %

t − 1 to t + 1, t + 2, t + 3 periods respectively are regressed on acquisition dummy and year dummies. We also estimated the regression with percentage change of R&D intensity from t − 1 to the average R&D intensity of three post-acquisition years.

Focusing on the specification including only acquisition dummy, the results indicate no impact in t + 1 and t + 2 year on R&D intensity. In t + 3 year the results explained significantly negative impact of M&A on R&D intensity relative to non-acquirers. Taking the average of three post acquisition years, it is observed that acquirers experience a significantly lower R&D intensity on average relative to non-acquirers (the R&D intensity falls by one percentage point). The result of this study differs from previous studies like Haspeslagh and Jemison (1991) and Desyllas and Hughes (2010) who concluded significant negative effect on R&D intensity in first post-acquisition year and positive impact in third year. These results are similar to those of Ornaghi (2009) and Danzon et al. (2007). Even the changeover average of three years also confirms lowering of R&D intensity for M&A firms in comparison to non-M&A firms in post-acquisition years.

Discussing the impact of M&A on R&D intensity in the model including augmented specification, positive and statistically significant effect of acquisition is observed in t + 1 year. But statistically significant negative impact is observed for t + 2 and t + 3 years. Even for the average of three post acquisition years M&A has significant negative impact on acquiring firms' R&D intensity. Perhaps, it could be concluded that for the case of Indian pharmaceutical sector the result contradicts the idea that M&A deliver appropriate economies of scale and knowledge synergies. As explained by Desyllas and Hughes (2010) the use of propensity score has adjusted estimation of the causal acquisition effect on R&D intensity and eliminated the potential downward bias.

4.2.2 The Technology Import Effect on Acquirers R&D Intensity

Analyzing the impact of acquiring firms' technological imports on change in R&D intensity in post-acquisition years, the results suggest separate impact for disembodied technology imports and embodied technology imports. Disembodied technology imports of acquiring firms significantly lower the R&D intensity in all the three consecutive post-acquisition years as well as for average of three years. Embodied technology imports enhance in-house R&D expenditure of acquiring firm but the results are statistically significant only for t + 3 year and for the average of three years. The results clearly indicate that disembodied technology imports are substitute to the technology acquired through M&A and reduce R&D investments in post-acquisition years while embodied technology imports are complementary to technology acquired through M&A and encourage in-house R&D expenditure.

4.2.3 Financial Capacity of Acquirers and R&D Intensity

Turning towards the impact of acquirers' financial characteristics, a strong support for negative relationship between acquirers leverage level and post M&A R&D intensity is observed. Statistically significant negative relationship is confirmed between acquirer's leverage level and R&D intensity for t + 1, t + 3 and the average of three post-acquisition years. These results are similar to that of Hall (1990), Hitt et al. (1991, 1996). According to these researchers, high debt level will force firms to provide significant amount of cash flows to debt repayment leaving fewer funds for investment in R&D projects. The result explains that a unit increase in level of leverage in t + 1 year after acquisition will decline R&D intensity by 21percent and 15 percent for the average of three post acquisition years. It could be concluded that high leverage is associated with added controls by investors thus; compelling managers to avoid risky investment in R&D projects with long payback periods in post-acquisition period rather than in earlier loosely monitored situations. In contrast, no significant impact of leverage growth at the time of acquisition on R&D intensity is observed. In case of non- acquirers, a significantly positive impact of leverage level on R&D intensity is observed but the coefficient values are very low and stable for all the post-acquisition years. It is likely that non-acquirers might be taking debt to invest in R&D processes and maintain their competitive positions.

4.2.4 Control Variables and the Impact on R&D Intensity

The implication of control variables for R&D intensity is also explored in augmented specification model. The results indicate absence of statistically significant impact of cross border acquisitions on R&D intensity for first two years after acquisitions but found positive and statistically significant effect in third year and also for average of three years. This later effect would indicate that the absorption of new knowledge encourages acquirers to devote greater internal efforts to exploit the technological capabilities of the acquired firms (Martin and Alvarez 2009). Positive and significant coefficient for time invariant average of three years clearly confirm that cross border acquisitions boost R&D intensity of acquirers more than domestic deals. The relatedness of acquirers and targets is captured by horizontal acquisition dummy. Positive and significant impact of M&A on R&D intensity is observed in third post-acquisition year and also for the average of three post acquisition years. Immediate post acquisition years are consumed in restructuring and integration of target and acquirer firms. Therefore, in later years relatedness effect is prominent as duplication of assets is avoided and complementarity of technological assets of parties involved help in improving R&D efficiency of acquirers.

5 Conclusions

The paper attempts to address the question whether M&A in Indian pharmaceutical sector improves innovative performance of firms analyzed in terms of R&D intensity. We also examine the role of acquirer's characteristics in the form of absorptive and financial capacity in order to find whether some acquirers are more successful than others. The results over the three post acquisition years and the aggregate of the three years suggest positive to negative effects of acquisitions on R&D intensity. The findings of this study are similar to many of the studies conducted in European and American high technology industries. It could be concluded from the paper that M&A have negative impact on R&D intensity of pharmaceutical firms but the negativity reduces over the period of time. The analysis also points out that if the firms undertake related and cross-border acquisitions, positive impact on R&D intensity can be observed. Heterogeneity of the acquiring and target firm and cross-fertilization of ideas strengthened R&D activity of the firms in post-acquisition period.

The results of weighted least squares regression analysis suggest that acquisition bring about slight positive impact on R&D intensity in first year (model 2) followed by subsequent negative impact in t + 2, t + 3 years and in aggregate of three years. This finding is likely to reflect the immediate benefits exploited by acquirer firms from target firms' R&D capabilities. But the subsequent years reveal the influence of bureaucratic hurdles, restructuring cost, integration issues and disruption of established organizational and R&D routines in both target and acquirer firms causing depressing effects on R&D intensity (Ranft amd Lord, 2002). We found in the study that the magnitude of negative effect on R&D intensity of acquiring firm has diminished over time.

As suggested by the earlier studies (Miller 1990), level of leverage matters for the acquisition effect on R&D intensity. Our results confirm that leverage level tends to impact negatively the R&D intensity of acquiring firms but leverage growth at the time of acquisition is not having any significant effect on R&D intensity. The lag of leverage level suggests the possible debt financing of M&A. These results are similar to that of Hall (1990) but they differ from those of Desyllas and Hughes (2010). The huge amount of leverage disables acquiring firms to afford the necessary post acquisition R&D investments.

Some acquirers are in a superior position to carry out acquisitions due to their enhanced absorptive capacity [measured by size of acquirer's technological knowledge base represented by technological imports in this study]. R&D intensity is differently affected by technological imports of acquirers. Disembodied technology affects negatively the R&D intensity but embodied technology has significantly positive impact on R&D intensity of acquirers. M&A allows acquiring firms to acquire tacit knowledge thus lowering the need of disembodied technology imports. It is observed that embodied technology imports is complementary to technology acquired via M&A and boosts in-house R&D expenditure.

In case of related acquisitions we find neutral effect in first two post acquisition years but significantly positive impact in third post acquisition year as well as for average of three years. Integration of two firms takes time therefore; results of cross fertilization of common but not similar ideas and resource deployment between two firms can be visible in later years of acquisitions. Interrelatedness also promotes technological complementarities and closeness of ideas that lead to more potential knowledge spillover thus, promoting enhanced R&D investments.

Cross-borders deals also affect positively the R&D intensity in later years of acquisition. Cross border deals can generate distinct complementarity creating knowledge transfer across boundaries. The heterogeneity of M&A partners help firms in shaping their innovative capabilities (Bertrand and Zuniga 2006). Geographical distances and different enterprise culture makes assimilation and application of technology time consuming therefore, synergies could be realized in the post acquisition years.

Firms can opt for technology acquisition through imports or by forming strategic alliances, over and above their in-house R&D efforts, in order to revitalize their existing knowledge base. They can also try to overcome the inertia and technological exhaustion occurred through current exploitation of existing knowledge base (Vermeulen and Barkema 2001) through imports or acquisitions of technologically active entities [other firms or R&D laboratories]. Number of firms based in India is trying to use these options over the last few years, especially in the Pharmaceutical sector. The R&D intensity of firms in the Pharmaceutical sector has also been relatively higher than that of firms in other sectors in India. The possible positive impact of M&A per se on their in-house technological efforts appears to take longer time to materialize. A thorough investigation over a longer period of time is, therefore, recommended before drawing firmer conclusions. However, in the light of our results we suggest that to reap the benefits of technology acquisition, integration process with those acquired or merged should be carried out carefully and effectively.

Acknowledgments An earlier version of this paper was presented in the 9th Annual Conference of the Knowledge Forum held during October 2014. The authors are grateful to Professors Siddharthan and Goldar for their valuable feedback. The comment and suggestions received from other participants are also highly appreciated. The error(s) that remain are our own.

Appendix A

See Table 3.

Table 3 Logit regression for estimating the propensity score of acquisition (Dependent Variable = Acquisition $_t$)

Variables	Coefficient estimates
Constant	−5.44 (−8.25)***
SIZE	0.59 (7.39)***
DTI	−51.42 (−1.68)*
LEV	−0.00009(−0.01)
ETI	1.43 (0.62)
PROF	−0.015 (10.15)
DZERO R&D	−0.79 (−2.74)***
RDI	0.91 (1.30)
GROWTH	0.0005 (0.06)
Time dummies	Yes
No. of observations	1360
L R χ^2 (15)	185.19***
Log likelihood	−345.10
Pseudo R^2	0.2116

*** Indicates 1 %

* Indicates 10 %

Appendix B

Reduction of mean bias after matching

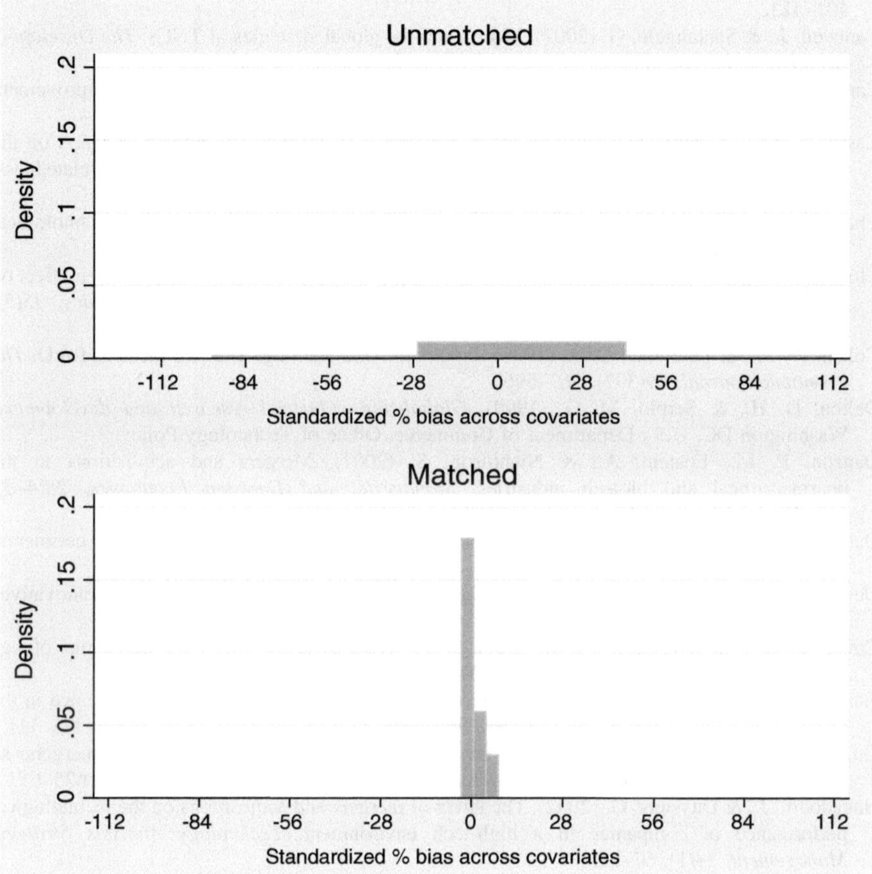

References

Ahuja, G., & Katila, R. (2001). Technological acquisitions and the innovation performance of acquiring firms: a longitudinal study. *Strategic Management Journal, 22*(3), 197–220.

Anand, J., & Singh, H. (1997). Asset redeployment, acquisitions, and corporate strategy in declining industries. *Strategic Management Journal, 18*(Special Issue), 99–118.

Barkema, H. G., & Vermeulen, F. (1998). International expansion through start-up or acquisition: A learning perspective. *Academy of Management Journal, 41*(1), 7–26.

Barney, J. B. (1991). Firm resources and sustained competitive advantage. *Journal of Management, 17*(1), 99–120.

Bertrand, O. (2009). Effect of foreign acquisition on R&D activity: Evidence from firm level data of France. *Research Policy, 38*(6), 1021–1031.

Bertrand, O., & Zuniga, P. (2006). R&D and M&A: are cross-border M&As different? An investigation of OECD countries. *International Journal of Industrial Organization, 24*(2), 401–423.

Cantwell, J., & Santangelo, G. (2002). M&As and the global strategies of TNCs. *The Developing Economies, 40*, 400–434.

Capron, L., & Mitchell, W. (1998). Bilateral resource redeployment and capabilities improvement following horizontal acquisitions. *Industrial and Corporate Change, 7*(3), 453–484.

Cassiman, B., Colombo, M., Garonne, P., & Veugelers, R. (2005). The impact of M&A on the R&D process—An empirical analysis of the role of technological and market relatedness. *Research Policy, 34*(2), 195–220.

Chakrabarti, A., Hauschildt, J., & Suverkrup, C. (1994). Does it pay to acquire technological firms? *R&D Management, 24*(1), 47–56.

Cloodt, M., Hagedoorn, J., & Van Kranenburg, H. (2006). Mergers and acquisitions: their effect on the innovative performance of companies in high-tech industries. *Research Policy, 35*(5), 642–654.

Cohen, W. M., & Levinthal, D. A. (1989). Innovation and learning: The two faces of R&D. *The Economic Journal, 99*(397), 569–596.

Dalton, D. H., & Serpio, M. G. (1999). *Globalizing industrial research and development.* Washington DC, U.S.: Department of Commerce, Office of Technology Policy.

Danzon, P. M., Epstein, A., & Nicholson, S. (2007). Mergers and acquisitions in the pharmaceutical and biotech industries. *Managerial and Decision Economics, 28*(4–5), 307–328.

Dehejia, R. H., & Wahba, S. (2002). Propensity score matching methods for non-experimental causal studies. *The Review of Economics and Statistics, 84*(1), 151–161.

Desyllas, P., & Hughes, A. (2010). Do high technology acquirers become more innovative? *Research Policy, 39*(8), 1105–1121.

Ernst, H., & Vitt, J. (2000). The influence of corporate acquisitions on the behaviour of key inventors. *R&D Management, 30*(2), 105–119.

Gantumur, T., & Stephan, A. (2007). Mergers & acquisitions and innovation performance in the telecommunications equipment industry. Electronic Working Paper Series, CESIS, No. 111.

Gugler, K., Mueller, D. C., Yurtoglu, B. B., & Zulehner, C. (2003). The effects of mergers: an international comparison. *International Journal of Industrial Organization, 21*(5), 625–653.

Hagedoorn, J., & Duysters, G. (2002). The effect of mergers and acquisitions on the technological performance of companies in a high-tech environment. *Technology Analysis Strategic Management, 14*(1), 67–85.

Hall, B. H. (1990). *The impact of corporate restructuring on industrial research and development.* Special issue on Microeconomics: Brookings Papers on Economic Activity.

Haspeslagh, P. C., & Jemison, D. B. (1991). *Managing acquisitions: creating value through corporate renewal.* New York: Free Press.

Healy, P. M., Palepu, K. G., & Ruback, R. S. (1992). Does corporate performance improve after mergers? *Journal of Financial Economics, 31*(2), 135–176.

Heckman, J., Ichimura, H., & Todd, P. (1997). Matching as an econometric evaluation estimator: evidence from evaluating a job training programme. *Review of Economic Studies, 64*(4), 605–654.

Henderson, R., & Cockburn, I. (1996). Scale, scope and spillovers: The determinants of research productivity in drug discovery. *Rand Journal of Economics, 27*(1), 32–59.

Hennart, J. F., & Park, Y. R. (1993). Greenfield vs. acquisitions: the strategies of Japanese investors in United States. *Management Science, 39*, 1054–1070.

Hirano, K., Imbens, G. W., & Ridder, G. (2003). Efficient estimation of average treatment effects using the estimated propensity score. *Econometrica, 71*(4), 1161–1189.

Hitt, M., Hoskisson, R., Ireland, R. D., & Harrison, J. (1991). Are acquisitions: A poison pill for innovation? *Academy of Management Journal, 34*(3), 693–706.
Hitt, M., Hoskisson, R., Johnson, R. A., & Moesel, D. D. (1996). The market for corporate control and firm innovation. *Academy of Management Journal, 39*(5), 1084–1119.
Huber, G. P. (1991). Organization learning: The contribution processes and the literatures. *Organization Science, 2*(1), 88–115.
Jensen, M. C. (1986). Agency costs of free cash flow, corporate finance and takeovers. *American Economic Review, 76*(2), 323–329.
Kogut, B., & Zander, U. (1992). Knowledge of the firm, combinative capabilities, and the replication of technology. *Organization Science, 3*(3), 383–397.
Makadok, R. (2001). Towarda synthesis of the resource-based and dynamic-capability views of rent creation. *Strategic Management Journal, 22*(5), 387–401.
Markides, C. M., & Ittner, C. D. (1994). Shareholder benefits from corporate international diversification: Evidence from US International Acquisitions. *Journal of International Business Studies, 25*(2), 343–366.
Martin, R., & Alvarez, I. (2009). Technological effects of M&As in Spanish manufacturing. *Industrial and Corporate Change, 18*(4), 761–784.
Miller, R. B. (1990). Do merger and acquisitions hurt R&D? *Research Technology Management,* 11–14.
Myers, S. C., & Maljuf, N. S. (1984). Corporate financing and investment decisions when firms have information that investors do not have. *Journal of Financial Economics, 13*(2), 187–221.
Narayanan, K. (1998). Technology acquisition, de-regulation and competitiveness: A study of Indian automobile industry. *Research Policy, 27*(2), 215–228.
Ornaghi, C. (2009). Mergers and innovation in big pharma. *International Journal of Industrial Organization, 27*(1), 70–79.
Prabhu, J. C., Chandy, R. K., & Ellis, M. (2005). The impact of acquisitions on innovation: poison pill, placebo, or tonic? *Journal of Marketing, 69*(1), 114–130.
Ranft, A. L., & Lord, M. D. (2002). Acquiringnewtechnologies and capabilities: a grounded model of acquisition implementation. *Organization Science, 13*(4), 420–441.
Ravenscraft, D. J., & Scherer, F. M. (1987). *Mergers*. Brookings Institution, Washington: Sell-offs and Economic Efficiency.
Rosenbaum, P. R., & Rubin, D. B. (1983). The central role of the propensity score in observational studies for causal effects. *Biometrika, 70*(1), 41–55.
Smith, C. L., & Warner, J. B. (1979). Bankruptcy, secured debt, and optimal capital structure: A comment. *Journal of Finance, 34*(1), 247–251.
Teece, D., Pisano, G. P., & Shuen, A. (1997). Dynamic capabilities and strategic management. *Strategic Management Journal, 18*(7), 509–533.
Vermeulen, F., & Barkema, H. (2001). Learning through acquisitions. *Academy of Management Journal, 44*(3), 457–476.
Veugelers, R. (2008). Literature review in mergers & acquisitions. . In B. Cassiman & M. G. Colombo (Eds.), *The innovation impact*. New York, USA: Edward Elgar.
Wagner, M. (2011). Acquisition as a means for external technology sourcing: Complementary, substitutive or both? *Journal of Engineering and Technology Management, 28*, 283–299.

Part II
Social Dimensions

Innovation and Employment: A Firm Level Study of Indian Industries

Arup Mitra and Amit Kumar Jha

1 Introduction

Innovation is endemic to economic growth. Growth in output which is more than proportionate increase in inputs is attributed to total factor productivity growth (TFPG), after controlling for returns to scale. TFPG in turn is a combination of technological progress and change in technical efficiency and an important determinant of total factor productivity growth, particularly the technological progress component, is linked to innovation. Endogenous growth models urge that research and development (R&D) expenditures taken as a broad proxy for innovative moves contribute directly to firms' productivity enhancement, and indirectly through their industry-wide spill-over effects (see Grossman and Helpman 1990; Romer 1986).

However, the effect of innovation on employment is an important concern: whether innovation tends to reduce employment or it can be conducive to employment growth is a pertinent issue. If innovation means lesser utilisatiion of all the factors of production for the same level of output to be produced, then naturally it tends to reduce employment per unit of output. However, if innovation reduces the utilization of some of the factors of production and not labour, then both innovation and employment can go hand in hand. In support of this latter view it

Article Note The financial support received under the IDRC-TTI grant to pursue this research is gratefully acknowledged.

This chapter was earlier published as "Innovation and employment: a firm level study of Indian industries" in the *Eurasian Business Review*, Vol. 5, No. 1, 2015 and is reproduced here with permission.

A. Mitra (✉) · A.K. Jha
Institute of Economic Growth, Delhi University, Delhi 110007, India
e-mail: arup@iegindia.org

A.K. Jha
e-mail: amit_jha612@rediffmail.com

N.S. Siddharthan and K. Narayanan (eds.), *Technology*,
India Studies in Business and Economics,
DOI 10.1007/978-981-10-1684-4_7

may be argued that output growth can be faster than the growth of some of the inputs such as capital but not labour because the labour contracts may involve rigidity. Labour might have been hired on a long term basis, which can be treated as a sunk cost (fixed or variable)—a retrospective cost that has already been incurred and cannot be recovered. Besides, the operation of the new technology is not necessarily automated that involves labour displacement.

A related point is also of great interest. Even if innovation leads to lesser utilization of all the factors (including labour) for a given level of output, the rise in the quantum of production certainly contributes to employment generation, i.e., the scale effect. Modernisation of technology may lead to its large scale application in various sectors of the economy and hence, the quantum of production and employment both may increase simultaneously even when the new technology gets more capital intensive. These issues of employment increase at the aggregate level due to wider application of the advanced technology prompted by the profit motive are certainly of great relevance, particularly in the context of the developing economies confronted with the compulsion of maximizing growth and generating employment opportunities for the vast supplies of labour.

However, there can be is a negative effect of innovation on employment. Since technological innovation largely takes place in developed countries they are made to suit these economies and their factor endowments. Incidentally these countries are primarily labour scarce and thus the new technology tends to become increasingly labour saving (Pack and Todaro 1969). Import of such technology by the developing countries reduces their employment growth, particularly in the high productivity formal sector. However, there could be a counter-argument to this as well. The inter-linkages between the formal and the informal sectors are of great significance. Ancillary activities which are undertaken in the informal sector may contribute to substantial employment generation and thus, at the aggregate level there could be employment gain. But this view can again be contested by arguing that employment in the informal sector is of residual type, which raises the vulnerability of the workers. On the whole, a thorough review of literature both on theoretical and empirical aspects needs to be pursued before turning to the empirical analysis.

2 Various Viewpoints: Existing Studies

Schumpeter (1939, 1961) initiated the concept of "innovation". In his postulation innovation is a new production function, displaying a new combination of factors of production or production conditions. Innovation is a continuous process of creative destruction, old being replaced by the new. The combination of capital, labor and other factors of production is optimized in the process of innovation and its impact on total employment and employment structure is cyclical. In the initial stages, total employment grows sluggishly or even declines, while employment structure does not change significantly; at a later stage, there are rapid increases in total employment and marked changes in employment structure; and in the final stage of innovation, changes both in total employment and employment structure gradually diminish until the next innovation comes through (Guangrong and Yuanyuan 2009).

If the new technology enhances productivity as well as promotes employment, the choice is clear. Such a possibility, though empirically difficult to materialize, exists at least theoretically. For example, technological progress brings in upward shift in the production frontier, which would mean higher levels of output for the given levels of inputs. In such a situation if the new technology becomes labour intensive, the rise in value added and employment both will be witnessed. However, the value added growth will be more than the rise in employment, and hence, labour productivity can actually shoot up.[1] Conversely, the new technology can dampen employment and improve productivity by adopting capital-deepening process.

Choi et al. (2002) analyze the implications of Hicks-neutral technical progress for a small Harris-Todaro economy with variable returns to scale. The analysis demonstrates that the welfare effects of technical progress consist of three components, i.e., the primary growth effect, the returns-to-scale effect and the employment effect. This type of decomposition is indeed useful as it deciphers the effects of technical progress into various components. Besides, the study works out the possibilities under non-constant returns to scale which is a much stronger possibility in the real world than a constant returns to scale situation. Under constant returns to scale the possibility of non-immiserizing exists and one may conclude that technical progress will be beneficial. But with the introduction of non-constant returns to scale, technical progress can lead to the returns-to-scale effect, which can be of any sign, and the sum of the primary growth effect and the employment effect again can be of any sign which creates the possibility of immiserizing growth. In other words, growth without employment generation is possible as technical progress tends to reduce labour absorption.

Technical progress and rising capital intensity in the literature are almost synonymous. On the other hand, innovations in the line of labour intensive technical progress is a difficult proposition as these innovations are not easy to pursue. The capital intensive technical change also has important implications for rates of industrialization and capital accumulation even when the economies, particularly in the developing world, are characterised by a dual economic structure. Kelley et al. (1972) noted that increases in the bias may tend to inhibit the rate of industrialization and reduce the rate of capital accumulation without appreciable changes in per capita GNP growth. Related to these results is the extent to which labour absorption in the industrial sector is affected: the study observes an important retarding influence that accumulates over time. It questions the wisdom of introducing labour saving

[1]However, when output is fixed, the shift in technology from being capital intensive to labour intensive would result in deterioration in labour productivity.

technology in the industrial sector in order to enhance per capita growth. The authors rather note that per capita income is mostly insensitive to the technological bias introduced in the industrial sector of the developing countries. Hence, the outcome is neither an increase in per capital income nor a rise in employment in the industrial sector in response to adoption of capital intensive technology.

In fact, Mureithi (1974) elucidates this point with great lucidity. The rising capital-labour ratio means that each job creation is becoming more capital-expensive. Of course it must not be supposed that rising capital intensity is bad per se. It is likely that a large part of the capital formation could be devoted to the building of infrastructure like roads, public works, communications, etc. In addition, as the author argues, it is pertinent to realize that production actually takes place in stages: (l) material handling, (2) material processing, (3) material handling among processes, (4) packaging, (5) storage of the finished products. Of the five stages, only the second, i.e., the central processing, need be capital intensive because at this stage the finer precision of temperature, pressure, ingredients combination, etc., is important. But there are many other stages where factor substitutability is technically possible and thus the entrepreneurs have a choice to select the technology. The desirability of a technology has to be judged not merely by its scientific or technical sophistication, but rather by its appropriateness in the context of the society in which it will be used. It requires innovative ideas to reduce the labour-saving elements of a technology while maintaining or improving quality and efficiency. In other words, labour-intensive technological progress which can improve performance and employment both is something that needs to be pursued by the developing countries. Even after accounting for the fact that there could be stages where capital intensive technology is absolutely necessary, innovation and employment can move in a positive direction in many other stages which then can offset the negative effect on employment as conceived in certain specific stages.

The "compensation theory" as Vivarelli points out, argues that technological unemployment is a temporary phenomenon. The labour saving effects of technology can be offset through: "(1) additional employment in the capital goods sector where new machines are being produced, (2) decreases in prices resulting from lower production costs on account of technological innovations, (3) new investments made using extra profits due to technological change, (4) decreases in wages resulting from price adjustment mechanisms and leading to higher levels of employment, (5) increases in income resulting from redistribution of gains from innovation, and (6) new products created using new technologies". [2]

Another interesting point emerges in relation to the preference for new technology (James 1993). If new technology is not adopted it may affect the quality of products and thus exports may suffer, resulting in employment loss. On the other hand, adoption of new technology which is capital intensive in nature can cause employment to fall. Hence, one has to verify if employment loss due to drop in exports prompted by the traditional labour intensive technology or employment loss

[2]Also see Vivarelli (1995) and Pianta (2005).

due to adoption of capital intensive technology is greater in magnitude. Further, the speed of production, product flexibility and locational factors need to be considered in assessing the total effect of technology on employment. If certain products are manufactured in the low cost countries labour intensive technology can still be pursued. Hence, the factor price ratio is an important determinant of technology choice and decision about location of production base, which eventually impact on employment. The idea of enlarging the production base across the globe is embed in the study by James (1993). While the low labour cost countries can specialize in the production of certain goods or certain components of the composite goods using the labour intensive methods, the developed countries may specialize in certain other components that require very high levels of capital and skill. Thus, the newer and innovative ways would mean that technical progress would not only suit the labour market situation of the developing and the developed countries both but also bring in a positive relationship between innovation and employment at large.

A positive relationship between innovation and employment has been conceptualized in a novel way by Saviotti and Pyka (2004). Interpreting economic development as synonym for new goods, services or sectors Saviotti and Pyka (2004) view their emergence as a result of increasingly systematic use of innovation—a component of economic development which amounts to a process of qualitative change within the economic system. It is quite natural that as the old product or services matures employability declines. This could be due to a number of reasons. For example, with learning by doing total factor productivity growth takes place and hence, to produce the same magnitude of goods/services labour requirement may decline sizably. Thus, to improve the level of employment in a continuous manner, innovation has to go on and new goods and services have to be produced. In this sense innovation and employment can go hand in hand. The ability to reap variety is a manifestation of economic development, which in turn can create employment steadily. Also, on the productivity front its growth may not take place indefinitely implying upper bounds on sectoral productivity growth. In order to augment the productivity growth at the country level efforts have to be pursued to create new sectors. On the whole, the possibility of a positive relationship between innovation, employment and growth is not inconceivable.

However, having said this we may turn to another set of argument which suggests that the developed countries are faced with a severe shortage of labour ready to pursue mechanical jobs, and thus the innovations relating to technical progress are usually pursued with an objective of reduction in labour requirement in the production process. So technical progress and rising capital intensity proceed synonymously, which do not conflict with the labour market situation in the developed countries. However, with import liberalization if the developing countries import this type of technology at a cheaper cost it restricts their employment growth. Thus the labor-saving technical change is a definite disadvantage to developing economies (Kelley et al. 1972).

As Azeez (2006) points out distinctly, a new technology gets embodied in capital goods, and therefore, import of capital goods is often considered as import of technology. Once imported capital good is put into operation, the technological progress realized in the country of origin will be incorporated into the production process (UNIDO 2005). Though UNIDO (2005) argues that it is still cheaper for a latecomer to buy the technology already invented by others than to re-invent the wheel, there can be serious implications in terms of employment loss. Chakravarty (1987) noted that with imports of capital goods on a significant scale, domestic costs of production are unlikely to come down since developing countries might be importing expensive capital goods. Further, imports of capital goods can act as substitutes for domestic production of capital goods, imposing a social cost in the form of unutilized capacity making the domestic firms operate at high unit cost of production (Chandrasekhar 1992).

The adoption and adaptation of these international technologies are indeed costly because of tacit knowledge and circumstantial sensitivity of technology (Evenson and Westphal 1995). Unless an importing country has significant technological capability, it cannot fully utilize the imported technology. Besides, imported technology may require more skilled than unskilled workers while developing countries are usually have an abundant supply of the latter type. Acemoglu and Zilibotti (2001) argue that due to the difference in skill scarcity, technology in developed countries tends to be skill intensive and is inappropriate for developing countries. Thus the potential productivity of imported technology cannot be realized in developing countries.

Next, one may pose the question in relation to product and process innovation. The interaction between economic integration, product and process innovation, and relative skill demand is an important aspect, which Braun (2008) analyses in a model of international oligopoly. Lowering of trade barriers increase the degree of foreign competition which may have effects on the incentives of firms to undertake R&D investment and also the firms' demand for skilled relative to unskilled workers. Increased competition following economic integration induces firms to bring down production costs by investing more aggressively in process R&D. At the same time, competitors expand their investments in product innovation in order to reduce the substitutability of their products. However, all this would require highly skilled human labour which can initiate newer ways of introducing cost efficient production processes and bring down the product differentials between the imported goods and the domestically produced goods. On the whole, economic integration and innovation are inter-linked resulting in an increase in the relative demand for skilled workers[3] and not the unskilled or semi-skilled variety of labour force which is in excess supply in most of the developing countries. Innovation and skill intensity usually go together—hence, even if innovation is not always labour displacing it benefits only those who are relatively in short supply. This tends to indicate that wage inequality is likely to increase in the process of innovation and increased trade.

On the empirical front Berman and Machin (2000) showed the skill-bias of technological change especially in middle-income countries. Pianta (2005)

[3]Vivarelli (2011) argues that innovation has a strong skill-bias.

emphasizes that innovation-based growth and job creation may operate in drastically different ways during different phases of the cycle, implying that the employment dynamics are not affected by the same factors and in the same ways during the upswings and the downswings. Piva (2003) presents a critical comparison of the positive implications of technology transfers (such as positive spillovers, technological catching-up, growing complementarities with domestic firms) with the negative ones (displacement of workers, negative welfare implications, competitive effects with domestic firms). Also, the author considers the nature of transferred technologies (labour-saving and/or skill-bias, embodied or not embodied in capital), together with the different institutional 'absorptive capacities' and sectoral specializations of both middle-income and low-income developing countries. Lee and Vivarelli (2006) suggest that import of capital goods may imply an increase in inequality via skill-biased technological change. Imports of capital goods,—embodying technological innovations—are important both because of the role they play in contributing to capital upgrading and more generally to the economic growth of the developing countries. In fact, even without necessarily assuming that developed countries transfer their "best" technologies, transferred technologies are relatively skill-intensive, i.e. more skill-intensive than those in use domestically before trade and FDI liberalization. Thus openness—via technology—should imply a counter-effect to the SS theorem prediction, namely an increase in the demand for skilled labour, an increase in wage dispersion and so an increase in income inequality. However, using data on 33 Indian manufacturing industries in India for the period, 1992 through 2001, Pandit and Siddharthan (2006) further showed that technology imports, through joint ventures and MNE participation, influence employment positively. They noted that employment growth, production of differentiated products, skill intensity of the work force and technological up-gradation go hand in hand. On the other hand, Mitra (2009) observed a decline in employment to value added ratio with a rise in manufacturing imports including technology.

Castellani and Zanfei (2006) present an in-depth theoretical and empirical analysis of the key issues underpinning the relationship between innovation and multinational companies. The authors argue that neither every foreign firm is a good source of externality nor every domestic firm is equally well placed to benefit from multinationals. Spillovers from multinationals differ according to the technological profiles, embedded-ness and linkage creation of both foreign and domestic firms. Hasan (2002) presented evidence from panel data on Indian manufacturing firms in favour of a significant effect of imported technology on productivity. The empirical literature on R&D, using cross-sectional data, reports strong evidence in favour of its positive effect on productivity while the time series estimates are less conclusive (Crespi and Pianta 2006).

With this background the rest of the paper is organized as follows. In Sect. 3 we estimate technical efficiency (TE) and total factor productivity growth (TFPG) of firms using panel data for each of the eleven manufacturing industries and Sect. 4 examines the impact of RND on some of these indicators. The impact of locational factor has also been controlled for. In Sect. 5 we assess the impact of R&D and other firm specific factors on employment. This is pursued with and without considering the simultaneous impact of efficiency and productivity on employment. Finally the findings are summarized in Sect. 6.

Two working hypotheses are pertinent:

1. Innovation leads to technology creation with higher efficiency and TFPG.
2. The productivity gains could be more in relation to capital than labour as the former is more expensive, which means innovation and employment both are positively associated.

Alternately, one may argue that expenditure on R&D has no effect on technology up-gradation since technology is mostly imported from the western developed countries. But this can be further contested on the ground that innovation is pursued by firms to attain product efficiency (quality), process other by-products and create ancillary products without any proportionate rise in investment in capital. This may require additional labour and thus the total employment of the firm may increase.

Using the firm level data in the manufacturing sector, compiled by ACCEQUITY for the period 1998 through 2010, the estimation has been carried out for eleven industries. The number of firms in many of these industries is substantial covering most of the existing ones with the exception of the very small ones.[4] However, the panel is not balanced as the information on all the variables is not available for all the firms and for each of the years. As regards the TFPG estimation the possibility of a strong relationship between TFPG and other inputs is low because we have considered an output function in which material and energy other than capital and labour (which are included in a value added function) are also considered. In other words TFPG estimated from a value added function may have strong relationship with inputs compared to TFPG estimated from an output function.

3 Estimating TE and TFPG

It may be useful at the outset to place a precautionary note in relation to the research and development expenditure (R&D) of the Indian firms, which is used as a proxy for innovation expenditure. First of all it does not necessarily mean technology creation. Funds are shown under R&D expenditure to derive tax benefits. After acquiring most parts of the technology from abroad firms have a tendency to show them as their own innovations. Similarly foreign personnel are invited to undertake R&D related activities rather than being pursued by the regular Indian employees.

[4]The number of firms is as follows: Consumer Durables (Domestic Appliances): 15, Consumer Durables (Electronics): 12, Chemical: 119, Electric Equipment: 51, Electronics Component: 36, Engineering: 79, Engineering Construction: 46, Engineering (Industrial Equipments): 38, Household and Personal Products: 23, Leather: 18, Pharmaceuticals and Drugs: 158.

The employees are sent abroad for training which is again then indicated under the broad head of R&D expenditure. Indian enterprises import technology from abroad through the employment of consultants (Mani 2008, 2009). In India, in-house R&D expenditures have increased tremendously since 1991 and so also the importation of technologies from abroad. The informal channels of technology imports have become very pronounced during the period of liberalization (Mani 2008, 2009).

Though the use of R&D is taken as a proxy for innovation it might be weak, especially in developing countries. Even if some MNEs have located their R&D headquarters in developing countries, computationally it does not take into account all the transfer of technologies embodied in machines which generally accounts for the most of the innovation dynamics. No information is available in this context from the companies' annual reports.

As Mani (2009) points out there is a strong view that India has now become a growing destination for innovative activities by MNCs, showing growing presence of foreign R&D centres in the country. Also FDI from India is steadily increasing and there have been a number of high profile take-over of Western technology-based companies by Indian corporates. All these indicators tend to suggest that India has become more innovative since 1991. Also, the TFPG growth in manufacturing and services is attributed to innovation. However this sort of an optimistic picture is an exaggeration. Though the technology oriented ventures are on the rise, Mani (2008, 2009) points out that most of the US patents that are assigned to India are actually owned by MNCs. The IT sector is well-known for obtaining patents but only foreign enterprises are specializing in IT related patents. On the whole, foreigners have taken more patents in India than Indians at the India Patent Office. He observed that the private sector enterprises are the most important inventors and among the ones who are active in obtaining patents are in the pharmaceutical sector. In fact, most of the domestic patents are in chemicals and pharmaceuticals only. So the view that India is becoming more innovative is rather difficult to be substantiated with evidence. Rather there is a severe shortage of skilled manpower, particularly in the Engineering sector (Mani 2008, 2009).

A critical review of the company reports also tends to support some of these views as they spend mostly for improvement in product quality by refining operational parameters in order to meet consumers' aspirations. Also they pursue processing of by-products, introduce new products and improve the existing product efficiencies. Technology absorption and adaptation, improvement in marketability of products with improvement in the quality and flexibility of the products are some of the important items. How R&D can help negate the effects of squeezed margins in the competitive markets in terms of new models, new versions, business growth is an important objective of the firms. However, the reports also mention the possibility of upgrading the technology and development of alternative low cost raw materials and optimization of energy consumption. Table 1 on R&D expenditure as a percentage of sales suggests that most of the industries correspond to a very low magnitude. Only pharmaceutical and drugs show a figure of 2 %; in the

Table 1 R&D as a percentage of sales (%)

Industry	RND/Sales (%)
Engineering (construction)	0.0073
Engineering	0.0309
Leather	0.0324
Chemical	0.0549
Consumer durables (electronics)	0.0624
Electronic components	0.0636
Engineering industrial equipment	0.11645
Consumer durables (domestic appliance)	0.17181
Electrical equipment	0.2547
Household and personal products	0.2653
Pharmaceutical and drugs	2.1651

rest of the industries it is substantially lower than 1 %. In terms of rate of growth, however, R&D expansion is enormously large which could be due to a low initial base.[5]

3.1 Time Variant TE and TFPG from Firm Specific Panel Data: Methodology

Total factor productivity growth encompasses technological progress (regress) and the change in technical efficiency (TE) over time. Technological progress may be defined as advances in knowledge relating to the art of production, which may take the form of new goods, processes or new modes of organization (Goldar 1986). On the other hand, technical efficiency is the efficiency with which factors of production are combined to generate output. Technological change can be conceptualized in terms of shifts in the production function (Solow 1957), whereas technical efficiency measures the distance between the actual and the frontier or maximum attainable levels of output (Bettesse 1990). Some of the earlier studies that estimated the total factor productivity growth in the Indian context did so mostly under the assumption of constant return to scale and perfect competition in the factor market (Goldar 1986). Without essentially considering these assumptions, the present study, based on the time series and cross-section poled data for fifteen major states in India, employs a frontier production approach to estimate technological change and the time varying technical efficiency, on the basis of which total factor productivity growth for each of the 17 two-digit level industries is estimated.

The methodology followed here is that of Cornwell et al. (1990), and has been applied in other studies as well (Krishna and Sahota 1991; Fecher and Pestiean 1993; Wu 1995).

[5]As for example, Whirlpool of India Ltd. registered a R&D growth of 28.41 %.

As mentioned above the firm level data is extracted from ACCEQUITY. Since we do not have information on output/value added, total sales have been taken as a proxy. The log of sales is regressed on log of employment, and expenditure on capital, material and fuel and a time trend (t), the coefficient of which captures the technological progress (regress). Since companies reported only wages and salaries bill and not the number of employees, the industry specific emolument per person has been taken from the Annual Survey of Industries to divide the wages and salaries bill and deduce the number of total employees, inclusive of floor workers.[6] Consumer price index for the industrial workers is used to deflate the nominal wages. Capital (gross asset) is deflated by the prices of machinery and machinery tools. Energy is deflated by prices of methanol. Sales and materials (raw) are deflated by the respective industry specific product prices (wholesale price index—WPI).

The equation for gross sales is estimated for each of the eleven industries applying the standard panel data methods. From the three versions of the model (OLS, FE and RE) the appropriate one is chosen on the basis of the Lagrange Multiplier statistic and the Hausman statistic. The results are presented in Table 2: most of the variables are statistically significant and among the significant ones all have positive coefficients.

In the next step the residuals are regressed on time t, i.e., time trend and t^2 for each of the company separately and the estimated values of the regress and are derived. Then pooling the data across firms over time the best performing firm and the year in which its performance is seen at the top is identified. This helps us estimate the efficiency level of the best performer for all other years and of all other firms for all the years. From the time varying technical efficiency and TFPG estimates at the firm level we have calculated the average estimates for each firm based on which the industry averages have been derived (Table 3). Consumer Durables (Domestic Appliances) and Engineering Construction are the two industries in which the average efficiency is over 60 %. Leather shows an efficiency of a little below 50 % and Consumer Durables (Electronics) and Engineering (Industrial Equipment) are at around 30 % level. Chemical, Electronics Component and Household and Personal Products are also close to 30 %. The rest of the three industries, i.e., Electric Equipment, Engineering, and Pharmaceutical, show a poor efficiency level of around 10 % or so. However, in Consumer Durables (Electronics) the average TFP growth has been exceptionally high (22.6 %) notwithstanding a moderate technical efficiency. Except the other three industries namely, Electric Equipment, Engineering, Engineering (Industrial Equipment) which experienced a TFPG of around 3–4 % the rest shows a sluggish growth of less than 1 % per annum.

[6]This procedure assumes that each firm in a given industry is paying the same wage rate which may not be the case in reality.

Table 2 Regression of sales on various inputs

Dependent variable—lnSales

Industry	Model	lnAssets	lnMaterial	lnEnergy	lnEmployment	Time	Constant	R^2	N
Consumer durables-domestic appliances	FE	−0.036 (−1.35)	0.787** (45.50)	0.073 ** (4.37)	0.169** (8.14)	0.008 ** (3.78)	1.684** (7.64)	0.99	214
Consumer durables-electronics	FE	0.076 (1.34)	0.632** (21.55)	0.036 (1.01)	0.351** (5.80)	0.004 (1.04)	1.307** (3.05)	0.96	175
Chemical	FE	0.059** (4.05)	0.699** (65.35)	0.125** (10.98)	0.080** (5.05)	0.003** (2.22)	1.634** (13.00)	0.96	1579
Electric equipment	FE	−0.042 (−1.32)	0.559** (28.02)	0.025 (1.13)	0.257** (7.96)	0.042** (10.30)	3.749** (13.60)	0.95	624
Electronics component	RE	0.044 (1.43)	0.696** (32.85)	0.059** (2.28)	0.252** (6.95)	0.012** (3.41)	1.274** (5.82)	0.97	473
Engineering	RE	0.153** (6.02)	0.285** (20.83)	0.158** (7.12)	0.354** (7.12)	0.040** (10.79)	2.612** (13.71)	0.90	1010
Engineering construction	RE	0.997** (288.74)	0.0003 (0.16)	0.002 (1.33)	−0.001 (−0.55)	0.013** (18.45)	−0.029* (−1.74)	0.99	463
Engineering-industrial equipments	FE	0.063 (2.78)	0.612 (30.39)	0.115 (5.55)	0.120 (4.71)	0.026 (7.01)	2.321** (14.28)	0.96	485
Household and personal products	FE	0.139** (4.24)	0.512** (14.57)	0.149** (5.04)	0.284** (7.36)	0.004 (1.17)	1.372** (4.32)	0.98	296
Leather	FE	−0.049 (−0.95)	0.758** (29.58)	0.164** (5.35)	0.044 (1.35)	0.010** (2.63)	2.022** (4.69)	0.97	249
Pharmaceuticals and drugs	FE	0.046** (2.89)	0.591** (65.82)	0.105** (8.41)	0.140** (9.58)	0.008** (4.21)	2.466** (20.64)	0.96	2061

ln stands for logarithmic transformation

Figures in parenthesis are t-values for FE model and z-value for RE model

FE denotes fixed effect model and RE denotes random effect model. Based on LM and Hausman statistics the appropriateness of the model is chosen. ** and * denote 5 and 10 % level of significance, respectively

4 Impact of R&D on TE and TFPG

The impact of R&D on performance, particularly TFPG, is a well-documented view. Nadiri (1993) found a positive and strong relationship between R&D as an index of technological change, and TFPG in the developed countries' context. Singh et al. (1996) also noted a positive impact of expenditures on basic, applied and experimental research on TFPG in Japan, South Korea and Taiwan. However, we noted in the text earlier that RND expenditure may not be always linked to technological change. In the next step we make an attempt to assess the impact of R&D as a percentage of sales on TE and TFPG. This is pursued in two ways. First, based on the firm-specific average figures the impact of R&D to sales ratio is examined after controlling for locational characteristic such as population base of the city/town where the firm is located. This is pursued keeping in view the literature on agglomeration economies (for a detailed review see Mitra 1999). Second, based on the panel data the R&D to sales ratio is taken to measure its impact on time varying-firm specific TE and TFPG. In this regression the effect of locational characteristic could not be considered due to the lack of time series data.

On an average employment in most of the industries seems to be expensive in terms of capital requirement as the employment per 100 thousand rupees of sales turns out to be quite low. In relative sense it is high, out of eleven industries, only in four, such as Leather, Engineering Construction, Electronic Components and Pharmaceutical and Drugs (Table 4).

The regression results show that the R&D to sales does not have a positive impact either on technical efficiency or TFPG in any of the industries considered in our analysis (Tables 5 and 6). Rather TFPG in Chemical is negatively related to

Table 3 Average efficiency and total factor productivity growth (%)

Industry	Efficiency	TFPG (%)
Consumer durables-domestic appliances	0.6	0.808
Consumer durables-electronics	0.384	22.576
Chemical	0.262	0.613
Electronics component	0.279	0.328
Electric equipment	0.096	4.098
Engineering construction	0.880	1.175
Engineering	0.123	3.847
Engineering-industrial equipment	0.344	2.667
Household and personal Products	0.289	0.408
Leather	0.479	1.19
Pharmaceutical	0.101	0.74

Source Authors' calculation

Table 4 Employment per sales (measured in 100,000 Rs)

Size classes	Industry
0–10	Chemical
10–20	Consumer durables (domestic appliances) Chemical Household and personal products Electrical equipment
20–30	Consumer durables (electronics) Engineering (industry equipment) Engineering
30–40	Pharmaceutical and drugs
40–50	Leather Engineering construction Electronic components

Table 5 Average technical efficiency and R&D/sales

Dependent variable—average efficiency: OLS					
Industry	Avg RND/Sales	Population	Constant	Adj-R sqr	N
Consumer durables-domestic appliances	24.897 (1.79)	1.07e−08 (1.01)	0.460** (5.77)	0.13	9
Consumer durables-electronics	−59.364 (−1.33)	6.31e−09 (0.37)	0.533** (5.39)	0.13	5
Chemical	−0.875 (−0.63)	3.96e−10 (0.32)	0.250** (26.56)	−0.02	66
Electric equipment	−1.010 (−0.64)	−1.48e−09 (−1.04)	0.095** (7.50)	−0.02	32
Electronics component	2.312 (0.76)	−1.18e−09 (−0.22)	0.258** (6.97)	−0.11	15
Engineering	−14.499 (−1.66)	−1.32e−09 (−0.30)	0.180** (4.73)	0.02	27
Engineering construction	0.258 (0.30)	2.60e−11 (0.03)	0.878** (108.44)	−0.15	15
Engineering-industrial equipments	−2.523 (−0.51)	8.34e−09 * (1.87)	0.346** (12.55)	0.19	13
Household and personal products	−4.930 (−1.64)	−2.78e−09 (−0.74)	0.329** (9.32)	0.05	13
Leather	−34.693 (−1.01)	−9.40e−09 (−0.27)	0.562** (4.90)	−0.17	7
Pharmaceuticals and drugs	0.008 (0.38)	1.26e−09* (1.87)	0.092** (18.80)	0.01	118

Note Figure in parenthesis are t-values. ** and * denote 5 and 10 % level of significance, respectively

Table 6 Average TFPG and R&D/sales

Dependent variable—average TFPG: OLS estimates					
Industry	RND/Sales	Population	Constant	Adj-R sqr	N
Consumer durables-domestic appliances	0.267 (0.16)	−1.10e−09 (−0.84)	0.0110 (1.13)	−0.13	9
Consumer durables-electronics	159.700 (0.75)	−6.86e−08 (−0.84)	0.0470 (0.10)	−0.45	5
Chemical	−1.602** (−3.03)	−3.32e−10 (−0.71)	0.01** (2.79)	0.11	66
Electric equipment	−3.25 (−1.78)	1.53e−09 (0.93)	0.050** (3.48)	0.08	32
Electronics component	−1.80 (−0.70)	−2.05e−09 (−0.46)	0.036 (1.25)	−0.12	13
Engineering	1.28 (0.20)	−1.27e−09 (−0.45)	0.051** (2.28)	−0.06	29
Engineering construction	0.226 (0.37)	−2.31e−10 (−0.39)	0.011** (2.18)	−0.09	19
Engineering-industrial equipments	−2.429 (−1.00)	−4.21e −09* (−2.13)	0.042** (2.94)	0.21	13
Household and personal products	−0.223 (−0.18)	−1.22e−09 (−0.78)	0.013 (0.92)	−0.13	13
Leather	−8.112 (−1.57)	−1.10e−08 (−2.08)	0.043* (2.54)	0.30	7
Pharmaceuticals and drugs	−0.0071545 (−0.27)	−1.33e−10 (−0.15)	0.009228 (1.53)	−0.0165	118

Note Figure in parenthesis are t-values. ** and * denote 5 and 10 % level of significance, respectively

R&D. Similarly the agglomeration variable also does not indicate any significant effect which could be due to the fact that we have taken the firms registration office address in the absence of plant address. Only two industries namely Engineering (industrial equipments) and Pharmaceutical and Drugs show a positive impact of population size on efficiency.

However, based on the panel data we are able to see a positive effect of R&D to sales ratio on technical efficiency in four industries (Chemical, Engineering (Industrial Equipments), Leather, Pharmaceuticals and Drugs), a negative effect in three other industries (Consumer durables-electronics, Electric equipment, Electronics component), and a statistical insignificance in the rest of the four groups (Table 7). In the case of TFPG the evidence is even scanty—two groups show a positive effect, i.e., Consumer Durables (Domestic Appliances), Engineering (Industrial Equipments), one negative, i.e., Pharmaceutical, and the rest are statistically insignificant (Table 8). Incidentally the Pharmaceutical industry reported the

Table 7 Technical efficiency and R&D/sales

Dependent variable—efficiency, panel data					
Industry	Model	RND/Sales	Constant	R^2/Adj R^2	N
Consumer durables-domestic appliances	RE	8.065** (5.91)	0.550** (16.15)	0.23	78
Consumer durables-electronics	OLS	−27.054** (−4.18)	0.453** (20.53)	0.36	30
Chemical	RE	0.684* (1.87)	0.243** (37.42)	0.004	586
Electric equipment	RE	−0.422** (−2.54)	0.084** (12.16)	0.003	220
Electronics component	RE	−1.669* (−1.81)	0.271** (12.45)	0.0001	103
Engineering	RE	0.361 (0.22)	0.142** (7.13)	0.03	185
Engineering construction	OLS	−0.114 (−0.03)	0.878** (186.99)	−0.01	83
Engineering-industrial equipments	RE	1.482** (2.85)	0.334** (12.25)	0.02	97
Household and personal products	RE	−0.475 (−0.35)	0.279** (20.67)	0.02	64
Leather	RE	2.765* (1.78)	0.491** (17.62)	0.03	54
Pharmaceuticals and drugs	RE	0.165** (6.76)	0.096** (30.21)	0.04	1176

Note Figure in parenthesis are t-values for FE model and OLS and z-value for RE model. ** and * denote 5 and 10 % level of significance, respectively. *FE* denotes fixed effect model: *RE* denotes random effect model: *OLS* denotes ordinary least square: *Adj* R^2 is calculated only for OLS

maximum innovation (as noted by Mani 2008 and also in Table 1). The negative coefficient in the equation for TFPG could be because of the fact that R&D expenditure is actually incurred to improve product quality etc. which has nothing to do with technological progress as such. Rather too much of experimentation may reduce performance in terms of TFPG. However, the extent of the available technology used, which is measured in term of technical efficiency, is positively associated with RND in this industry. This differential impact of RND on TE and TFPG becomes very distinct as we change the regressor to log of R&D instead of R&D to sales ratio. The positive effect of ln R&D on TE is noted in a number of industries whereas the negative effect is evident in the case of TFPG (Tables 8 and 9). This is indicative of the fact that given the technology firms are able to improve the performance (TE) by spending more on R&D. However, R&D is not able to contribute to overall TFPG as technology is often imported from abroad. We have not controlled for other variables in these equations mainly to examine the gross relationship between R&D and the performance indicator and also because of the lack of information on variables which may be relevant from the point of view of

Table 8 TFPG and R&D/Sales

Dependent variable—TFPG; panel data					
Industry	Model	RND/Sales	Constant	R^2/Adj R^2	N
Consumer durables-domestic appliances	RE	3.292** (6.37)	−0.004 (−0.77)	0.17	78
Consumer durables-electronics	OLS	−32.55 (−0.64)	0.506** (2.90)	−0.02	30
Chemical	RE	0.025 (0.06)	0.008 (1.57)	0.005	586
Electric equipment	RE	−0.766 (−1.06)	0.062** (4.68)	0.0001	220
Electronics component	RE	2.013 (1.60)	−0.0006 (−0.04)	0.01	99
Engineering	RE	0.163 (0.07)	0.050** (2.72)	0.001	185
Engineering construction	OLS	−3.517 (−1.06)	0.011** (2.42)	0.001	83
Engineering-industrial equipments	RE	1.320** (2.68)	0.010 (1.16)	0.003	97
Household and personal products	RE	1.950 (1.21)	−0.009 (−0.75)	0.04	64
Leather	RE	−1.145 (−1.07)	0.022** (2.56)	0.02	54
Pharmaceuticals and drugs	RE	−0.425** (−6.10)	0.012** (2.69)	0.03	1176

Note Figure in parenthesis are t-values for FE model and OLS and z-value for RE model. ** and * denote 5 and 10 % level of significance, respectively. *FE* denotes fixed effect model: *RE* denotes random effect model: *OLS* denotes ordinary least square: *Adj* R^2 is calculated only for OLS

Table 9 Technical efficiency and lnR&D

Dependent variable—efficiency: panel data					
Industry	Model	lnRND	Constant	R^2/Adj R^2	N
Consumer durables-domestic appliances	RE	0.023** (6.56)	0.484** (13.52)	0.36	78
Consumer durables-electronics	RE	−0.021** (−4.76)	0.465** (10.05)	0.46	30
Chemical	RE	0.007** (6.39)	0.219** (29.62)	0.11	585
Electric equipment	RE	0.002** (4.36)	0.069** (12.52)	0.55	220
Electronics component	RE	0.0003 (0.08)	0.261** (10.39)	0.0002	101
Engineering	RE	0.018** (5.16)	0.093** (4.14)	0.02	184
Engineering construction	OLS	0.002	0.866**	0.01	80

(continued)

Table 9 (continued)

Dependent variable—efficiency: panel data					
Industry	Model	lnRND	Constant	R^2/Adj R^2	N
		(1.36)	(96.09)		
Engineering-industrial equipments	FE	0.007** (3.66)	0.307** (30.63)	0.33	96
Household and personal products	FE	0.0104** (2.27)	0.194** (6.78)	0.17	60
Leather	RE	0.017** (2.63)	0.432** (13.97)	0.39	54
Pharmaceuticals and drugs	FE	0.003** (7.66)	0.079** (28.24)	0.33	1170

Note Figure in parenthesis are t-values for FE model and OLS and z-value for RE model. ** and * denote 5 and 10 % level of significance, respectively. *FE* denotes fixed effect model: *RE* denotes random effect model: *OLS* denotes ordinary least square: *Adj* R^2 is calculated only for OLS

the firm.[7] A similar attempt is also made in the paper by Singh and Trieu (1996) (Table 10).

5 Impact of R&D on Employment

Employment to sales ratio perceived as a rough proxy for labour requirement per unit of output has been regressed on R&D to sales ratio, exports to sales ratio, imports to sales ratio, assets to sales ratio and efficiency (or TFPG). In an alternative specification employment to sales ratio has been replaced by log of employment, without changing the determinants. This is pursued mainly to capture the view that labour per unit of real output (approximated by real sales) may not increase in response to R&D though the overall employment may.[8] The performance indicator is included to test if TFP growth, for example, results in higher output growth relative to input growth including labour or alternately employment does not drop though the use of other factor inputs may decline.

In the equations with technical efficiency as one of the determinants the following three industry groups unravel a positive effect of R&D to sales ratio on employment: Engineering (Industrial Equipment), Household and Personal

[7]Though it is an omitted variable model it clearly indicates the lack of a relationship. If the impact were significant then there was a reason to consider other variables before highlighting the result.

[8]If the rise in output is more than employment then labour per unit of output may decline in spite of an increase in overall employment.

Table 10 TFPG and ln R&D

Dependent variable—TFPG: panel data					
Industry	Model	lnRND	Constant	R^2/Adj R^2	N
Consumer durables-domestic appliances	FE	0.008** (5.09)	−0.022** (−3.98)	0.15	78
Consumer durables-electronics	OLS	−0.002 (−0.08)	0.429* (1.90)	−0.03	30
Chemical	FE	−0.007** (−4.44)	0.034** (5.39)	0.02	585
Electric equipment	RE	−0.0004 (−0.18)	0.061** (3.43)	0.0003	220
Electronics component	FE	0.012 (1.41)	−0.018 (−0.72)	0.11	97
Engineering	RE	−0.010* (−1.93)	0.080** (3.37)	0.03	184
Engineering construction	OLS	−0.004** (−2.18)	0.025** (2.86)	0.04	80
Engineering-industrial equipments	FE	−0.007** (−3.74)	0.056** (5.77)	0.001	96
Household and personal products	RE	0.0001 (0.03)	−0.008 (−0.38)	0.04	60
Leather	FE	−0.013** (−2.77)	0.0707** (3.72)	0.001	54
Pharmaceuticals and drugs	FE	−0.005** (−3.34)	0.035** (3.93)	0.0003	1170

Note Figure in parenthesis are t-values for FE model and OLS and z-value for RE model. ** and * denote 5 and 10 % level of significance, respectively. *FE* denotes fixed effect model: *RE* denotes random effect model: *OLS* denotes ordinary least square: *Adj* R^2 is calculated only for OLS

Products, Pharmaceutical and Drugs (Table 11). In the rest of the industries R&D to sales ratio remains insignificant. Technical efficiency itself shows a negative effect on employment to sales ratio in the case of Electronics Component and a positive effect in engineering (industrial equipments) and remains insignificant in the rest of the industries.

The ratio of exports to sales is significant with a positive coefficient in three industries and negative only in one. Similarly the imports to sales ratio show a significant value only in three industries and among them two are positive. Based on this it is difficult to generalize that trade contributes to employment generation. However, some of the labour intensive sectors like Consumer Durables (Domestic Appliances) and Household and Personal Products show a positive effect of both export to sales and import to sales on employment to sales. While higher exports lead to increased employment, imported inputs also tend to create employment, suggesting possibilities of complementary relationship between the imported inputs

Table 11 Employment/Sales and R&D/Sales with TE

Dependent variable: employment/sales

Industry	Model	RND/Sales	Export/Sales	Import/Sales	Asset/Sales	TE	Constant	R^2/Adj R^2	N
Consumer durables-domestic appliances	OLS	−60.906 (−0.33)	5.501* (1.96)	35.868** (4.03)	1.452 (0.40)	0.988 (0.15)	7.307 (1.61)	0.37	29
Consumer durables-electronics	OLS	309.098 (1.55)	−32.785** (−2.91)	3.074 (0.98)	−10.741 (−1.77)	21.641 (1.49)	2.868 (0.50)	0.70	18
Chemical	RE	30.674 (1.23)	−1.045 (−1.00)	−0.837 (−0.66)	1.856** (4.44)	1.002 (0.24)	3.063** (2.36)	0.07	186
Electric equipment	RE	203.766 (1.22)	−1.135 (−0.22)	−0.448 (−0.05)	1.484 (0.81)	−31.656 (−1.14)	9.350** (2.86)	0.03	96
Electronics component	OLS	31.550 (0.18)	32.149** (2.94)	19.948 (1.66)	0.172 (0.64)	−83.524** (−3.53)	29.970** (4.19)	0.41	32
Engineering	RE	161.433 (0.49)	−1.057 (−0.19)	−2.679 (−0.42)	8.776** (3.37)	−21.047 (−1.22)	10.140** (2.68)	0.40	45
Engineering construction	OLS	3968.547 (1.21)	−4299.384 (−1.45)	17.053 (0.82)	0.687 (0.10)	−40.983 (−1.12)	40.997 (1.20)	0.36	7
Engineering-industrial equipments	OLS	1431.894** (2.62)	2.444 (0.31)	−91.238** (−3.37)	8.080** (3.57)	104.567** (2.88)	−28.425** (−2.10)	0.68	31
Household and personal products	RE	374.736** (3.00)	18.092** (2.13)	41.933** (2.71)	2.105 (1.10)	−27.171 (−1.31)	14.116** (2.31)	0.02	46
Leather	RE	1799.993 (0.92)	15.840 (1.62)	−30.546 (−0.93)	−5.998 (−0.33)	−29.798 (−1.31)	27.965 (1.63)	0.31	26
Pharmaceuticals and drugs	FE	56.842** (5.53)	−3.520 (−1.37)	1.738 (0.42)	8.696** (318.97)	−41.859 (−1.45)	11.692** (3.33)	0.97	499

Note Figure in parenthesis are t-values for FE model and OLS and z-value for RE model. ** and * denote 5 and 10 % level of significance, respectively. *FE* denotes fixed effect model: *RE* denotes random effect model: *OLS* denotes ordinary least square. *Adj* R^2 is calculated only for OLS

and skilled labour. Not any major improvement in results is obtained by redefining the dependent variable as log transformation of employment.[9]

When we replace technical efficiency by TFPG in the equation for employment to real sales ratio, the results relating to RND/Sales ratio remain unchanged except for Electronics Component which now turns out to be negative and significant (Table 12). TFPG itself is significant only in two industries with a negative coefficient, implying higher growth in output relative to input growth. Electronics Component and Household and Personal Products show a positive effect of both exports and imports. Even after changing the dependent variable to log of employment both these industries continue to indicate the positive effect of trade. Also, after changing the dependent variable to log of employment Electronics Component, Engineering (Industrial Equipment) and Leather show a positive effect of R&D to sales on employment with no negative effect in any of the other industries.[10]

On the whole, the R&D/sales ratio is not significant in a number of industries; however the cases of positive impact are noteworthy.

In the light of the second hypothesis we have dropped TFPG or technical efficiency from the employment equation. The effect of R&D/sales on employment to sales turns out to be positive and significant in the following four industries: Electric Equipment, Engineering (Industrial Equipments), Household and Personal Products, Pharmaceuticals and Drugs. In the rest of the industries the effect is statistically insignificant. As we redefine the dependent variable in terms of log of employment, the R&D to sales ratio takes a positive coefficient for Consumer Durables (Domestic Appliances) and negative for Consumer Durables (Electronics).[11]

The export to sales ratio is positive in four industries and negative in two industries. Similarly the import to sales is positive in three industries and negative in only one. Interestingly all these three industries showing positive effect of imports, also show the positive effect of exports (e.g., Consumer durables-domestic appliances, Engineering construction, Household and personal products). In five industries the asset to sales ratio shows a positive effect on employment (Table 13).

In several studies employment is taken to be a function of value added and wage rate to estimate the growth and wage elasticity of employment. Following the same logic we may regress log of employment on log of real sales, real wage rate (derived by deflating the nominal figures by the consumer price index for industrial workers), and in addition real RND (deflated by the price index for machinery). Since R&D/Sales ratio has a highly limited variation across companies and over time, log of R&D may be considered to be more suitable.

In this specification (Table 14), log R&D turns out to be significant with a positive effect in a number of industries (seven) and the elasticity of employment

---

[9]Results not reported.

[10]Results not shown.

[11]Results not reported.

Table 12 Employment/Sales and R&D/Sales with TFPG

Dependent variable: employment/sales

Industry	Model	R&D/Sales	Export/Sales	Import/Sales	Asset/Sales	TFPG	Constant	R^2/Adj R^2	N
Consumer durables-domestic appliances	OLS	−49.604 (−0.26)	5.494 (1.68)	35.483** (4.04)	1.390 (0.38)	1.541 (0.05)	7.933** (4.69)	0.37	29
Consumer durables-electronics	OLS	119.961 (0.50)	−32.922* (−2.72)	−1.491 (−0.69)	−3.828 (−0.42)	−1.265 (−1.10)	10.801** (7.99)	0.67	18
Chemical	FE	25.810 (0.98)	−0.244 (−0.20)	−0.115 (−0.09)	1.964** (4.43)	−1.069 (−0.24)	3.216** (6.56)	0.03	186
Electric equipment	RE	171.269 (1.03)	0.343 (0.07)	−3.024 (−0.37)	2.482 (1.45)	4.331 (0.37)	6.066** (2.76)	0.005	96
Electronics component	RE	−425.871* (−1.92)	14.137* (1.94)	22.107** (2.24)	1.062** (3.22)	−49.693** (−2.88)	5.626** (2.79)	0.47	30
Engineering	RE	127.925 (0.38)	−1.050 (−0.19)	−5.082 (−0.78)	9.592** (3.92)	−14.093 (−1.08)	7.162** (2.73)	0.28	45
Engineering construction	OLS	−1633.149 (−0.24)	1167.036 (0.27)	32.055 (1.49)	7.146 (1.05)	28.362 (0.69)	1.882 (0.64)	0.02	7
Engineering-industrial equipments	OLS	2076.12** (3.56)	10.245 (1.16)	−76.600** (−2.45)	4.745* (1.74)	−5.985 (−0.08)	8.787* (1.79)	0.58	31
Household and personal products	RE	403.708** (2.88)	19.146** (2.07)	38.48** (2.23)	2.055 (1.01)	−9.267 (−0.81)	6.910** (3.48)	0.02	46
Leather	OLS	1065.961 (0.65)	22.450** (2.64)	−28.044 (−1.04)	9.830 (0.62)	−128.113** (−3.22)	5.005 (0.44)	0.38	26
Pharmaceuticals and drugs	RE	58.454** (5.93)	−5.212** (−2.29)	0.099 (0.03)	8.701** (323.13)	0.970 (0.10)	7.278** (5.09)	0.98	499

Note Figure in parenthesis are t-values for FE model and OLS and z-value for RE model. ** and * denote 5 and 10 % level of significance, respectively. *FE* denotes fixed effect model: *RE* denotes random effect model: *OLS* denotes ordinary least square. *Adj* R^2 is calculated only for OLS

Table 13 Employment/Sales and R&D/Sales without performance indicator

Dependent variable—employment/sales								
Industry	Model	RND/Sales	Export/Sales	Import/Sales	Asset/Sales	Constant	R^2/Adj R^2	N
Consumer durables-domestic appliances	OLS	−42.461 (−0.30)	5.834** (2.58)	36.029** (4.77)	0.684 (0.25)	8.011** (5.44)	0.41	30
Consumer durables-electronics	OLS	254.966 (1.24)	−38.581** (−3.49)	−0.612 (−0.30)	−11.041 (−1.74)	11.256** (8.67)	0.67	18
Chemical	FE	25.097 (0.96)	−0.267 (−0.22)	−0.085 (−0.07)	1.961** (4.44)	3.224** (6.61)	0.03	186
Electric equipment	RE	160.121 (0.98)	0.818 (0.17)	−4.441 (−0.59)	2.532 (1.57)	6.356** (3.17)	0.004	97
Electronics component	RE	390.092** (2.58)	39.079** (2.43)	13.932 (1.13)	0.311 (1.08)	4.041 (0.86)	0.17	32
Engineering	RE	167.599 (0.52)	−1.038 (−0.19)	−3.338 (−0.54)	10.296** (4.40)	6.249** (2.55)	0.28	47
Engineering construction	OLS	252.464 (1.28)	9.079* (2.49)	28.312** (2.92)	5.148 (1.58)	3.081** (3.20)	0.46	10
Engineering-industrial equipments	OLS	2068.919** (3.67)	10.132 (1.19)	−76.14** (−2.53)	4.861** (2.18)	8.708* (1.85)	0.60	31
Household and personal products	RE	355.182** (2.79)	15.870* (1.84)	44.558** (2.82)	2.594 (1.38)	6.597** (3.81)	0.03	47
Leather	FE	221.470 (0.25)	−0.415 (−0.05)	9.806 (0.54)	41.032** (4.75)	3.798 (0.46)	0.01	26
Pharmaceuticals and drugs	RE	124.543** (113.87)	−10.507** (−4.02)	−6.449 (−1.50)	8.706** (266.35)	8.473** (5.75)	0.98	507

Note Figure in parenthesis are t-values for FE model and OLS and z-value for RE model. ** and * denote 5 and 10 % level of significance, respectively. *FE* denotes fixed effect model: *RE* denotes random effect model: *OLS* denotes ordinary least square. *Adj* R^2 is calculated only for OLS

Table 14 Partial elasticity of employment with respect to sales, wages and R&D

Dependent variable—lnEmployment							
Industry	Model	lnSales	lnR&D	lnWagerate	Constant	R^2/Adj R^2	N
Consumer durables-domestic appliances	RE	0.784** (15.03)	0.051** (2.02)	−0.470 (−1.40)	−2.36** (−2.78)	0.89	78
Consumer durables-electronics	OLS	0.498** (8.74)	0.297** (8.48)	−0.325 (−0.41)	−0.448 (−0.35)	0.95	33
Chemical	FE	0.497** (14.99)	0.015 (0.74)	−0.745** (−8.59)	1.180** (4.13)	0.71	586
Electric equipment	FE	0.484** (10.34)	0.070** (3.98)	−0.821** (−4.61)	1.55** (4.49)	0.85	225
Electronics component	RE	0.581** (14.03)	0.066** (2.38)	−1.348** (−6.79)	1.358** (2.69)	0.92	101
Engineering	FE	0.477** (11.79)	0.023 (0.96)	−1.051** (−16.56)	2.712** (7.37)	0.77	186
Engineering construction	RE	0.825** (10.30)	0.007 (0.28)	−1.012** (−2.34)	−1.760** (−2.29)	0.73	83
Engineering-industrial equipments	RE	0.732** (17.16)	0.017 (0.67)	−0.861** (−8.28)	−0.559 (−1.40)	0.84	98
Household and personal products	RE	0.745** (12.07)	0.065** (2.62)	−0.904** (−5.30)	−0.476 (−0.70)	0.95	61
Leather	RE	0.802** (6.96)	0.153** (2.19)	0.124 (0.17)	−3.142** (−2.45)	0.67	54
Pharmaceuticals and drugs	FE	0.443** (18.30)	0.117** (7.67)	0.007 (0.07)	0.811** (3.55)	0.82	1194

Note Figure in parenthesis are t-values for FE model and OLS and z-value for RE model. ** and * denote 5 and 10 % level of significance, respectively. *FE* denotes fixed effect model: *RE* denotes random effect model: *OLS* denotes ordinary least square. *Adj* R^2 is calculated only for OLS

with respect to R&D is seen to be highest in Consumer Durables (around 0.3). In two other industries (Leather and Pharmaceutical) it is again a little above 0.1. In Electric Equipment, Electronics Component and Household and Personal Products also the estimate is closer to 0.1.

6 Conclusion

On the whole, we noted that R&D as a percentage of sales does not affect efficiency or TFPG significantly. However, in absolute terms its impact on efficiency turns out to be positive in a number of cases though in terms of TFPG the effect is negative. This is indicative of the fact that given the technology firms are able to improve the performance (TE) by spending more on R&D. However, R&D is not able to contribute to overall TFPG as technology is often imported from abroad. After netting out the change in efficiency the TFP growth is attributed to technology up-gradation, which is sought from the developed countries. Hence, R&D expenditure does not enable firms to attain better performance relating to new technology procured from abroad—rather it reduces the overall performance (TFPG) possibly because of high adaptation cost of the new technology and inability to operate it and reap its potentiality instantaneously.

The impact of R&D as a percentage of sales on employment is positive only in a few industries. This has been tested with and without controlling for the performance indicator, which does not show any strong effect on employment. The R&D to sales ratio is seen to have a very limited variation for which the log transformation of R&D has been tried to work out the partial elasticity of employment with respect to R&D. In this specification a number of industries reported a positive effect of R&D on employment in absolute sense. Also, some of the labour intensive industries revealed a higher elasticity of employment with respect to R&D. On the whole, R&D's positive impact on employment in absolute sense if not employment content per unit of output, is noteworthy even when R&D does not mean actual innovation. Processing of byproducts and efforts pursued to bring in an improvement in product quality and efficiency are some of the striking features of R&D expenditure, which may be resulting in employment gains. Even when capital intensive technology is adopted by the firms R&D expenditure has the potentiality to generate employment as it means additional activities without involving additional capital.

References

Acemoglu, D., & Zilibotti, F. (2001). Productivity differences. *Quarterly Journal of Economics, 116*, 563–606.

Ahmed, I. (1987). Technology, production linkages and women's employment in South Asia. *International Labour Review, 126*, 21.

Ahmed, I. (1988). The bio-revolution in agriculture: Key to poverty alleviation in the third world? *International Labour Review, 127*, 53.

Azariadis, C., & Pissarides, C. (2007). Unemployment dynamics with international capital mobility. *European Economic Review, 51*, 27–48.

Azeez, A. E. (2006). Domestic capacity utilization and import of capital goods: Substitutes or complementary? Evidence from Indian capital goods sector. In D. S. Tendulkar et al. (Eds.) *India: Industrialisation in a reforming economy (essays for K.L. Krishna)*. New Delhi: Academic Foundation.

Berman, E., & Machin, S. (2000). Skill-biased technology transfer around the world. *Oxford Review of Economic Policy, 16*(3), 12–22.

Berman, E., & Machin, S. (2004). Globalization, skill-biased technological change and labour demand. In E. Lee & M. Vivarelli (Eds.), *Understanding globalization, employment and poverty reduction* (pp. 39–66). New York: Palgrave Macmillan.

Bettesse, G. E. (1990). Frontier Production Functions and Technical Efficiency: A Survey of Empirical Applications in Agricultural Economics. Paper No 50, Dept of Econometrics, University of New England, Australia.

Birla Institute of Scientific Research. (1980). Technological changes in agriculture: Impact on productivity and employment, Economic Research Division, New Delhi Vision Books.

Braun, S. (2008). Economic integration, process and product innovation, and relative skill demand. *Review of International Economics, 16*, 864–873.

Castellani, D., & Zanfei, A. (2006). *Multinational firms, innovation and productivity*. UK: Edward Elgar Cheltenham.

Caballero, R. J., & Hammour, M. L. (1996). On the timing and efficiency of creative destruction. *The Quarterly Journal of Economics, 111*, 805–852.

Cepede, M. (1972). The green revolution and employment. *International Labour Review, 105*, 1.

Chakravarty, S. (1987). *Development planning: The Indian experience*. New Delhi: Oxford University Press.

Chandrasekhar, C. P. (1992). Investment behaviour, economies of scale and efficiency in an import substituting regime: A study of two industries. In A. Ghosh, et al. (Eds.), *Indian industrialisation*. New Delhi: Oxford University Press, Structure and Policy Issues.

Choi, J.-Y., Yu, E. S. H., & Jin, J. C. (2002). Technical progress, urban unemployment, outputs, and welfare under variable returns to scale. *International Review of Economics & Finance, 11*, 411.

Cornwell, C., et al. (1990). Production frontiers with cross-sectional and time series variation in efficiency levela. *Journal of Econometrics, 46*(1–2), 185–200.

Crespi, Francesco and Mario Pianta. (2006). Demand and innovation in productivity growth. http://www.ec.unipg.it/DEFS/uploads/crespiantaproductivity.pdf.

Evenson, R., & Westphal, L. E. (1995). Technological change and technology strategy. Chap. 37, In J. Behrman, & T. N. Srinivasan, (Eds.), *Handbook of development economics* (vol. 3A, pp. 2209–2229), Amsterdam, North-Hollan.

Fecher, F., & Pestiean, P. (1993). Efficiency and competition in OECD financial services. In O. Harold, C. A. Fried, L. Knox, & S. S. Schmidt (Eds.), *The measurement of productivity efficiency, techniques and applications* (pp. 374–85), New York: Oxford University Press.

Goldar, B. N. (1986). *Productivity growth in Indian industry*. Delhi: Allied Publishers.

Garmany, J. W. (1978). Technology and employment in developing countries. *The Journal of Modern African Studies, 16*, 549–564.

Grossman, G. M., & Helpman, E. (1990). Trade, innovation and growth. *The American Economic Review, 80*,(2). In: Papers and proceedings of the hundred and second annual meeting of the American Economic Association, May, pp 86–91.

Tong, G., & Liu, Y. (2009). The Affection of independent innovation on employment. *Management Science and Engineering, 3*(1), 36–40.

Hasan, R. (2002). The impact of imported and domestic technologies on the productivity of firms: Panel data evidence from Indian manufacturing firms. *Journal of Development Economics, 69*, 23–49.

James, J. (1993). New technologies, employment and labour markets in developing countries. *Development and Change, 24*, 405.

Lee, E., & Vivarelli, M. (2006). *The social impact of globalisation in the developing countries*. CSGR Working Paper No 199/06, March.

Kelley, et al. (1972). Biased technological progress and labour force growth in a dualistic economy. *Quarterly Journal of Economics*, *86*, 426–447.

Krishna, K. L., & Sahota, G. (1991). Technical efficiency in Bangladesh manufacturing industries. *The Bangladesh Development Studies, 19*, 89–101.

Mani, S. (2008). *Financing of industrial innovations in India: How effective are tax incentives for R&D?* Centre for Development Studies, Working Paper No. 405.

Mani, S. (2009). *Has India become more innovative since 1991? Analysis of the evidence and some disquieting features*. Centre for Development Studies, Working Paper No. 415.

Mitra, A. (2009). Technology import and industrial employment: Evidence from developing countries'. *Labour, 23*(4), 697–718.

Mitra, A. (1999). Agglomeration economies as manifested in technical efficiency of firms. *Journal of Urban Economics, 45*(3), 490–500.

Mureithi, L. (1974). Demographic and technological variables in Kenya's employment scene. *Eastern Africa Economic Review, 6*, 27–43.

Nadiri, M. I. (1993). *Innovation and technological spillovers*. NBER Working Papers, 4423.

Pack, H., & Todaro, M. (1969). Technological transfer, labour absorption, and economic development. *Oxford Economic Papers, 21*, 395–403.

Pandit, B. L., & Siddharthan, N. S. (2006). MNEs, product differentiation, skills and employment: Lessons from the Indian experience. In S. R. Hashim & N. S. Siddharthan (Eds.), *High-tech industries, employment and global competitiveness* (pp. 165–179). New Delhi: Routledge.

Pianta, M. (2005). Innovation and employment. In J. Fagerberg, D. Mowery, & R. R. Nelson (Eds.), *Handbook of innovation* (pp. 568–598). Oxford: Oxford University Press.

Piva, M. (2003). The impact of technology transfer on employment and income distribution in developing countries: A survey of theoretical models and empirical studies. In *International Labour Office, Policy Integration Department, International Policy Group, Working Paper n. 15*, ILO, Geneva.

Romer, P. M. (1986). Increasing returns and long-run growth. *The Journal of PoliticalEconomy, 94,*1002–1037.

Saviotti, P. P., & Pyka, A. (2004). Economic development, variety and employment. *Revue Économique 55*, 1023–1049.

Schumpeter, J. A. (1939). *Business cycles: A theoretical, historical and statistical analysis of the capitalist process*. New York: McGraw-hill.

Schumpeter, J. A. (1961). *The theory of economic development*. New York: Oxford University Press.

Singh, I., & Richard, D. (1975). A microeconometric chronicle of the green revolution. *Economic Development and Cultural Change, 23*, 661.

Singh, N., & Trieu, H. (1996). *The role of R&D in explaining total factor productivity growth in Japan, South Korea, and Taiwan*. Santa Cruz: University of California. http://citeseerx.ist.psu.edu/viewdoc/download?doi=10.1.1.200.9362&rep=rep1&type=pdf

Singh, N., & Hung, T. (1996). *The role of R&D in explaining total factor productivity growth in Japan, South Korea, and Taiwan*. University of California, at Santa Cruz, http://citeseerx.ist.psu.edu/viewdoc/download?doi=10.1.1.200.9362&rep=rep1&type=pdf

Solow, R. M. (1957). Technical change and the aggregate production function. *Review of Economics and Statistics* 312–320.

UNIDO. (2005). *Productivity in developing countries: Trends and policies*. Vienna: United Nations Industrial Development Organisation.

Vivarelli, M. (1995). *The economics of technology and employment: Theory and empirical evidence.* Cheltenham: Elgar

Vivarelli, M. (2011). *Innovation, employment and skills in advanced and developing countries: A survey of the literature*. IDB Publications 61058, Inter-American Development Bank.

Wu, Y. (1995). Productivity growth, technological progress and technical efficiency change in China: A three sector analysis. *Journal of Comparative Economics, 21*(2), 207–229.

Impact of Agricultural Related Technology Adoption on Poverty: A Study of Select Households in Rural India

Santosh K. Sahu and Sukanya Das

1 Introduction

Growth in agricultural output is one of the most effective means to address poverty in the developing world. In this line of argument, the Department for International Development (2003) estimates that a 1 % increases in agricultural productivity could reduce poverty between 0.6 and 2 %. However, growing population is one of the major challenges in developing countries to increase agricultural productivity in a sustainable way, to meet the demand of the food security issues. The growth in production cannot come from area expansion but have to come from growth in yields emanating from scientific advances offered by biotechnology and other plant breeding initiatives (De Janvry et al. 2001). In the increasing research of improved varieties of major crops that enhanced the productivity of agriculture, impact assessment studies were conducted to arrive at the direct and the indirect welfare impacts. Kijima et al. (2008) in Uganda conducted a study on the impact of rice, and found that rice adoption reduces poverty without deteriorating the income distribution. Similarly, Winters et al. (1998), Mwabu et al. (2006), and Wu et al. (2010) show positive impact of agricultural technologies adoptions. However, there are serious complexities associated with understanding the impact pathways through which agricultural technology adoption might affect household welfare. This is because crop production can affect household welfare both directly and indirectly. Consistent

S.K. Sahu (✉)
Madras School of Economics, Gandhi Mandapam Road, Kottur, Chennai 600025, Tamil Nadu, India
e-mail: santosh@mse.ac.in

S. Das
Department of Policy Studies, Teri University, Plot no. 10, Vasant Kunj Institutional Area, New Delhi 110070, India
e-mail: sukanya.das@teriuniversity.ac.in

N.S. Siddharthan and K. Narayanan (eds.), *Technology*,
India Studies in Business and Economics,
DOI 10.1007/978-981-10-1684-4_8

with this notion, De Janvry et al. (2001) reports that crop production affects poverty directly by raising welfare of poor farmers who adopt the technological innovation, through increased production for home consumption, higher gross revenues from sales, lower production costs, and lower yield risks. The indirect ways through which crop production affects welfare include the prices of food for net buyers and employment and wage effects in agriculture and related activities.

In a poor and backward state like Odisha in India, it is least expected that the development scenario of the village and the pace of socio-economic transformations could be better. The state has to make a longitudinal perspective plan for the transformation of the subsistence oriented backward agricultural economy in order to solve the problem of poverty and to improve the 'quality of life' of rural people. Dependence of population on primary sector occupations is quite high, whereas agriculture with its present state of infrastructure and technology and, above all, operational holdings is itself not in a position to provide a substantial form of gainful livelihood to the majority of rural population in Odisha. There have been little occupational diversifications of population at the village level. Irrigation infrastructure created through many development projects has failed to achieve desired goal across space and people. Often this serves the interest of only large, medium and semi-medium farmers. Like this, the welfare programmes and Minimum Needs Programme implemented by the state under social sector development to lessen poverty and improve the 'quality of life' of poor in rural areas hardly reach the targeted sections of population.

In a poor and backward state showing highest incidence of rural poverty, the development of hardcore backward districts requires special attention. It is found that the socio-economic conditions of people in the undivided Kalahandi, Bolangir and Koraput districts, popularly known as the KBK districts, have worsened over the years. These three districts have been affected by the '*backwash effects*' of development, which may probably be due to their inherent disadvantageous factors like poor quality of human capital, backward and unsustainable agriculture coupled with reckless exploitation of forest resources. This study is an attempt to see the impact of agricultural technology adoption on poverty of the select households of the Jeypore sub-division of the Koraput district of Odisha state.

The undivided Koraput district is characterized by certain features, historical, natural and geographical. The district lies on a section of the Eastern Ghat discontinuous range of mountains and holds five natural divisions, with a mean elevation of 3000, 2500, 2000, 1000 and 500 ft above sea level, respectively. The district has two parts, each characterized by a distinct type of rock; the 2000 ft plateau of Jeypore, with its much lower extension down into the Malkangiri subdivision (Malkangiri district), and the high hilly regions of the Eastern Ghat, lying between the Jeypore plateau and the Visakhapatnam coastal plains. This geographical setting has to a large extent isolated the region from the plain coastal districts of Odisha. Among the consequences of this, the region has been able to preserve much of its varied and prolific fauna and flora, and its aboriginal inhabitants have not undergone radical change as a result of contact with modern civilization. The location of this area, economic activities and socioeconomic profile gives us the motivation of studying the impact of the agricultural technology adoption and its impact on poverty at household level.

The role of agricultural technology and its impact on rural poverty and fostering overall economic development has been widely documented in the economic literature. Although quite complex, the relationship between the adoption of new technology and poverty reduction has been perceived to be positive (Bellon et al. 2006; Binswanger and Von Braun 1991; Evenson and Gollin 2003; Just and Zilberman 1988). Productivity-improving technologies reduce poverty by reducing food prices, facilitating the growth of nonfarm sectors, and by stimulating the transition from low productivity subsistence agriculture to a high productivity agro-industrial economy (Just and Zilberman 1988). However, the potential for poverty reduction through reduced food prices, growth in the nonfarm sector and agricultural commercialization depends to a large extent on the magnitude of productivity gains in agriculture. However, the impact of agricultural technology adoption is necessary to understand at farm-household level. It is also important to distinguish between the direct and the indirect impact of the impact of such technology adoption (Becerril and Abdulai 2010; David and Otsuka 1994; De Janvry and Sadoulet 2002; Minten and Barrett 2008; Moyo et al. 2007).

The direct effects of new agricultural technology on poverty reduction are the productivity benefits enjoyed by the farmers adopting new technology. These benefits usually manifest themselves in the form of higher farm incomes. The indirect effects are productivity—induced benefits passed on to others by the adopters of the technology. These may comprise lower food prices, higher nonfarm employment levels or increases in consumption for all farmers (De Janvry and Sadoulet 2002). However, productivity—enhancing agricultural technology involves a bundle of innovations rather than just a single technology. The impacts of higher-order (indirect) benefits from technology adoption depend: depend on the elasticity of demand, outward shifts in supply lowering food prices; and an increased productivity which may stimulate the demand for labor. The poor and marginal farmers tend to supply off-farm labor, which may translate to increased employment, wages, and earnings for them. They have little or no land and they gain disproportionately from employment generated by agricultural growth and from lower food prices. Higher productivity can, therefore, stimulate broader development of the rural economy through general equilibrium and multiplier effects, which also contribute to poverty reduction. Agricultural technology may induce changes in cropping patterns and allocation of farmers' own resources to different uses. It is important to notice that the technology adoptions may vary from farmer to farmer and the nature of the technology in use. For instance, technology adoption in agriculture can be either through high yield variety (HYV) seeds, advances in irrigation facilities, fertilizers, pesticides use or through the machinery employed during agricultural activities.

Mendola (2007) adopts a non-experimental evaluation strategy in order to assess the direct contribution of modern-seed technology adoption to rural poverty in Bangladesh. Using a cross-sectional household survey from rural Bangladesh, the study isolates the causal effect of adopting high yielding varieties (HYVs) of rice on poverty alleviation by using the "propensity-score matching" (PSM) method. According to the PSM estimation method, the adoption of HYVs of rice has a positive impact on farm household wellbeing. Allowing for interactions between agricultural technology and other determinants of income, this method leads us to quantify the positive impact of technology adoption on resource-poor farmers, in terms of rise of income and poverty reduction.

Minten and Berrett (2008) study in Madagascar also drew similar conclusion of adopting of intensifying improved technologies which is strongly associated with better agricultural yields. Karanja et al. (2003) showed that maize technology adoption in high agricultural potential regions of Kenya is likely to have substantially greater positive impacts on aggregate real incomes, but may have a less-than-positive influence on income distributional outcomes, compared to technology adoption in low agricultural potential regions. Becerril and Abdulai (2010) also uses PSM to analyze the impact of the adoption of improved maize varieties on household income and poverty reduction, using cross-sectional data of 325 farmers in Mexico. The findings reveal a robust positive and significant impact of improved maize variety adoption on farm household welfare measured by per capita expenditure and poverty reduction. The adoption of improved maize varieties helped in raising the household per capita expenditure by an average of 136–173 Mexican pesos, thereby reducing their probability of falling below the poverty line by roughly 19–31 %.

Most of the studies on the impact of agricultural technology on farm incomes and poverty reduction focus macro approaches, with very few analyses at the micro-level. Some of the few household level studies include Evenson and Gollin (2003), Mendola (2007), and Moyo et al. (2007). Kassie et al. (2011) evaluates the ex-post impact of adopting improved groundnut varieties on crop income and poverty in rural Uganda. The study utilizes cross-sectional data of 927 households, collected in 2006, from seven districts in Uganda. Using PSM technique the study reports that adopting improved groundnut varieties (technology) significantly increases crop income and reduces poverty.

Thus, the literature appears to document overall positive impacts, with far less evidence at the individual household level that specifically show the effects of the adoption of agricultural technologies on farm productivity and household welfare. This study is a value addition in this regard in the context of Odisha. The objective of this paper is to assess the role of agriculture related technology adoption, on consumption expenditure and poverty status measured by headcount index, poverty gap index and poverty severity index. The empirical question that we would like to address is "*do agriculture related technology adoptions have the potential to reduce poverty?*" In understanding this question, we apply PSM method to deal with the selection bias problem. In addition to PSM, we also conduct the "*rbounds test*" and a "*balancing test*" using the "*mean absolute standardized bias*" between the agricultural technology adopters and non-adopters as suggested in Rosenbaum and Rubin (1985). The rest of the paper is organized as follows. Section 2 presents the analytical framework and the model, Sect. 3 presents the data and descriptive statistics, Sect. 4 presents the econometric results and Sect. 5 concludes.

2 The Analytical Framework

One of the standard problems in impact evaluation involves the inference of the causal relations between the treatment and the outcome. There are two specific related problems with regards to evaluating the impact, of an intervention on targeted individuals; such as (1) the selection bias problem and (2) missing data problem in case of the counterfactual. There is extensive literature describing developments in addressing the problem stated above. Broadly, empirical literature categorizes evaluation methods in five categories such as; (1) pure randomized experiments (2) natural experiment (3) matching method (4) selection or instrumental variable model and (5) structural simulation model. This paper aims at indentifying the causal effect of adoption of agricultural related technologies on consumption expenditure and poverty using matching method on the non-experimental data. We follow Imbens and Angrist (1994), using counterfactual outcomes framework known as the Average Treatment Effect (ATE). Under this framework, it is assumed that each observational experimental unit with an observed outcome has ex-ante two potential outcomes: (1) an outcome when under adoption (that we denote y_1) and (2) an outcome when not under adoption (we denote y_0).

Let y_i the *observed* overall expenditure for a household i. Thus y_1 and y_0 are two random variables representing, respectively, the potential expenditure level of household i when farmer participate in agriculture related technology adoptions $(d_i = 1)$ or does not participate $(d_i = 0)$. For any household i, the causal effect of participation in agriculture related technology adoption on household expenditure is defined as $(y_1 - y_0)$. However, the two potential outcomes can't be observed at the same time. We observe either y_1 or y_0. According to whether the household had participated or not, it is not possible to measure $(y_1 - y_0)$ directly. The average causal effect of adoption within a specific population (ATE) can be determined as $E(y_1 - y_0)$, with E as the mathematical expectation.

Several methods have been proposed to estimate ATE, and they include the matching methods based on propensity scores, as well as parametric methods based on Instrumental variable methods. The choice of method is largely driven by the assumptions made and the availability of data. For any observational data (that is non-experimental) an important assumption is; the Conditional Independence Assumption (CIA), that states conditional on X (observables), the outcomes are independent of the treatment (d) and can be written as:

$$y_{1,}y_0 \perp d|X \tag{1}$$

The behavioral implication of this assumption is that participation in the treatment does not depend on the outcomes after controlling for the variation in outcomes induced by differences in X. A much weaker assumption also used for *indentifiability* of the causal effect of the treatment is what Imbens and Angrist (1994) refers to as the *unconfoundedness* assumption, and which Rubin (1978) refers to as the *ignorability* assumption. The assumption is written as:

$$y_0 \perp d | X \tag{2}$$

If valid, the assumption implies that there is no omitted variable bias once X is included in the equation hence there will be no confounding. The assumption of *unconfoundedness* (Eq. 2) is very strong, and its plausibility heavily relies on the quality and the amount of information contained in X. A slightly weaker assumption also associated with the treatment effect evaluation is referred to as the "overlap or matching (common-support condition)" assumption. The assumption ensures that for each value of X, there are both treated and untreated cases. The assumption is expressed as follows:

$$0 < \Pr[d = 1|X] < 1 \tag{3}$$

This implies that there is an overlap between the treated and untreated samples. Stated the other way round this also means that the control and treated populations have comparable observed characteristics. Under the assumption discussed above (CIA and overlap) the ATE on the Average Treatment Treated (ATT) can be identified as:

$$\begin{aligned} E(y_1 - y_0 | a = 1) &= E[E(y_1 - y_0 | d = 1, X)] \\ &= E[E(y_1 | d = 1, X) - E(y_0 | d = 0, X) | d = 1] \end{aligned} \tag{4}$$

where, the outer expectation is over the distribution of X, in the subpopulation of participating households in agricultural related technologies. In observational data, it is not possible to calculate directly the difference in the outcome of interest between the treated and the control group or the ATE due to the absence of the counterfactual.[1] As a consequence, data may be drawn from comparison units whose characteristics match those of the treated group. The average outcome of the untreated matched group is assumed to identify the mean counterfactual outcome for the treated group in the absence of a treatment. The propensity score matching method matches treated and untreated cases on the propensity score rather than on the regressor. The propensity score which is the conditional probability of receiving treatment given X, is denoted $P(x)$ written as:

$$p(x) = \Pr[d = 1|X = x] \tag{5}$$

An assumption that plays an important role in treatment evaluation is the balancing condition which states that;

$$d \perp X | p(x) \tag{6}$$

[1]The counterfactual is a condition in which the same household is observed under treatment and without treatment. In reality a household can only be observed under either of the two conditions at a time and not under both.

This can be expressed alternatively by stating that, for individuals with the same propensity score the assignment to treatment is random and should look identical in terms of their x vector. The main purpose of the propensity score estimation is to balance the observed distribution of covariates across the groups of adopters and non-adopters (Lee 2005). The balancing test is normally required after matching to ascertain whether the differences in the covariates in the two groups in the matched sample have been eliminated, in which case, the matched comparison group can be considered a plausible counterfactual (Ali and Abdulai 2010). Although several versions of balancing tests exist in the literature, the most widely used is the mean absolute standardized bias (MASB) between adopters and non-adopters (Rosenbaum and Rubin 1985). Additionally, Sianesi (2004) proposed a comparison of the pseudo R^2 and p-values of the likelihood ratio test of the joint significance of all the regressors obtained from the logit analysis before and after matching the samples.

After matching, there should be no systematic differences in the distribution of covariates between the two groups. As a result, the pseudo R^2 should be lower and the joint significance of covariates should be rejected. Given how sensitive the quasi-experimental methods are to assumptions, we conduct the sensitivity analyses based on the Rosenbaum's method of sensitivity analysis as we assume CIA crucially depends on the possibility to match treated and control units on the basis of a large informative of pre-treatment variables. The threshold level of welfare that distinguishes poor households from non-poor households is the poverty line. Using a poverty line, a number of aggregate measures of poverty can be computed. A more general measure of poverty proposed by Foster et al. (1984) belongs to a class of poverty measures is given as:

$$p_\alpha = \frac{1}{n}\sum_{i=1}^{q}\left[\frac{z-y_i}{z}\right]^\alpha 1(y_i - z) \tag{7}$$

where, the poverty line is $(z), y$ is expenditure per capita of the ith household measured in the same unit as z, n is the total number of individuals in the population, q is total number of poor individuals whose income is less than the poverty line, $1(y_i - z)$ is indicator variable that takes value of one if the income is below the poverty line and 0 otherwise and a poverty aversion parameter that takes values of 0, 1, and 2, providing three commonly used indices of poverty; (1) poverty incidence as represented by the head count index, (2) intensity by the poverty gap index and (3) severity by the squared poverty gap index. The poverty line is a subsistence minimum expressed as in Rangarajan committee report (2014). Individuals who reside in households with consumption lower than the poverty line are then labeled "*poor*". Using the minimum food expenditure as an additional measure, we can identify the "*ultra poor*" households whose total consumption per capita on food and non-food items is lower than the minimum food expenditure.

3 Data and Descriptive Statistics

The data were collected through a household survey conducted in Koraput district of Odisha state in India. The sample villages are the beneficiaries of various programmes of M.S. Swaminathan Research Foundation (MSSRF) initiatives on technologies related to agriculture. The households were randomly selected from Jeypore sub-district. This led to the selection of 296 households. Data were collected at village and farm-household levels. At the village level, data collected included crops grown and the village infrastructures. At the household level data collected included the farmer knowledge of varieties cultivated, household composition and characteristics, land and non-land farm assets, livestock ownership, household membership to different rural institutions, varieties and area planted, indicators of access to infrastructure, household market participation, household income sources and consumption expenses. In this study, adopters are classified as households who have adopted at least one of the agricultural technologies, out of maximum of 17 technologies as reported by the sample households during the primary survey. These technologies are in terms of "asset related" to "technology related" suitable for agricultural activities such as use of tractors, motor for irrigation etc. weighted against the land holding (net). Table 1 reports descriptive statistics, disaggregated at the adoption status.

Table 1 presents a comparison of some of the important indicators at household level distinguished between adaptors and non-adopters. We can observe from the table that income, income less from MGNREGA, expenses related to food and total expenses, share of income from primary and secondary sources, are statistically significant between two groups. However, expenses related to non-food, income from tertiary source, age and education of head of households are not statistically different between both the groups. Therefore, determinants of poverty can be different or similar based on the variables that are statistically different. Further, we also know that there are trade-offs in technology that generates direct and indirect effects. When land is unequally distributed, and if there are market failures and conditions of access to public goods that vary with farm size, then the optimum farming systems will differ across farms. Small holder may opt to adopt capital saving technologies, while larger farmers may prefer capital intensive technologies.

Table 2 presents the distribution of sample households according to land holdings and adoption status. Consistent, with Becerril and Abdulai (2010), the differences in the distribution of land between adopters and non-adopters suggest a positive correlation between the incidence of adoption and the ownership of land. The incidence of adoption is clearly higher among 1st and 3rd quartiles of land distribution compared to the other two distributions. Such differences in land ownership between adopter and non-adopters could also contribute to the disparities in welfare indicators between the two groups.

Table 3 describes the comparative statistics between adopters and non-adopters household characteristics related to technology adoption and land holdings. Here, we have tried to compare the variables such as technology score, agriculture related

Table 1 Household characteristics by adoption status

Variables	Non-adopters (n = 107)	Adopters (n = 189)	Full sample (n = 296)	Difference (t-test)
Total income	37,297.680	48,143.480	44,222.870	2.495***
Income less from MGNREGA	36,422.920	47,202.210	43,305.640	2.468***
Food expenses	18,397.760	21,339.760	20,276.270	2.059***
Non-food expenses	5043.028	6289.159	5838.699	1.547
Total expenses	28,098.790	34,244.330	32,022.800	2.354***
Share of income from primary source	62.115	67.527	65.571	2.263***
Share of income from secondary source	23.773	19.806	21.240	2.362***
Share of income from tertiary source	6.311	5.621	5.870	0.683
Age of head of household	45.607	43.042	43.970	1.535
Education of head of household	0.645	0.630	0.635	0.261

Source Primary data collected by authors during 2014
Note *** indicate statistically significant at 1 %, MGNREGA—Mahatma Gandhi National Rural Employment Guarantee Act

Table 2 Distribution of sample households by landholding and adoption status

Quartile(s)	Non-adopters (n = 107)		Adopters (n = 189)	
	Frequency	Percentage	Frequency	Percentage
1st	16	14.95	44	23.28
2nd	21	19.63	31	16.40
3rd	23	21.50	67	35.45
4th	47	43.93	47	24.87

Source Primary data collected by authors during 2014

technology score, components of land in terms of total land, irrigated, dry, wet land and technology related expenses. These components or indicators are compared between the two groups. The technology score and the agriculture technology score are differentiated based on the technology related to agriculture and non agriculture. The score for each of the groups are defined as a weighted score that is similar to the Human Development Index (HDI).

From Table 3 we can observe that the sample, that is differentiated based on the adopters and non-adopters are statistically difference in terms of technology score, agriculture related technology score, ratio of wet land to total land and expenses related to technology at household level. Other than these variables, indicators such as ratio of dry land to total land, total land and ratio between irrigated and total land are not statistically different between two groups. Table 4 presents mean and median per capita consumption expenditure and the Gini coefficient by household grouped in different groups. There is a significant difference between the adopter categories in terms of welfare indictors. Table 4 also presents the estimated mean

Table 3 Comparative indicators households

Variables	Non-adopters (n = 107)	Adopters (n = 189)	Total	Difference (t-test)
Technology score	0.034	0.205	0.143	11.232***
Agriculture related technology score	0.001	0.298	0.191	17.680***
Wet land as a ratio of total land	0.216	0.278	0.256	1.844*
Technology related expenses	4658.000	6615.413	5907.834	1.660*
Dry land as a ratio of total land	0.143	0.161	0.155	0.686
Total land in acres	2.300	2.687	2.547	0.937
Irrigated land as a ratio of total land	0.426	0.444	0.438	0.405

Source Primary data collected by authors during 2014
Notes *** and * indicate statistically significant at 1 and 10 % level

Table 4 Mean and median per capita consumption expenditure, and the Gini coefficient

	Mean	Median	Gini coefficient
Male headed households	18,718.2	14,435.2	30.5
Female headed households	12,799.2	14,435.2	25.0
Household adopted to agricultural related technology	19,120.6	14,435.2	32.1
Household not adopted to agricultural related technology	17,324.8	14,435.2	28.1
Full sample	18,416.7	14,435.2	30.5

Source Primary data collected by authors during 2014

and median per capita consumption expenditure, and the Gini coefficient by household based on household head's characteristics. A further close look at the distribution of total expenses data shows that it is also skewed. After transforming the consumption variable into the logarithm form, the distribution is normalized but the t-test still shows a significant difference in consumption expenditure between adopters and non-adopters. About 91.37 % of the households live below the poverty line. The incidence of poverty is higher among the non-adopters (84 %) than it is among adopters (78 %) indicating an unconditional headcount ratio of poverty for the adopters of about 10 % points lower, compared to non-adopters. About 35 % of the households are ultra poor implying that 35 % of the households among the sample live in such dire poverty that they cannot even afford to meet the minimum standard of daily-recommended food requirement. The incidence of ultra poverty is also higher among non-adopters (46 %) than among adopters (39 %) suggesting that agriculture related technology adoption is positively correlated with wellbeing.

Table 5 presents the sensitivity of poverty measures to choice of indicator. This table gives the estimates of poverty headcount, poverty gap and squared poverty gap with and without some of the important indicators. For example, we can see

Table 5 Sensitivity of poverty measures to the choice of indicator

	Poverty headcount rate	Poverty gap	Squared poverty gap
Actual	91.4	53.2	34.3
Without technology adoption (absolute)	92.0	56.7	37.9
Without agricultural technology adoption (absolute)	93.5	59.7	41.1
With education	91.1	51.2	32.3
With technology score (relative)	89.7	49.1	30.3
With agricultural technology score (relative)	87.2	44.1	25.9

Source Primary data collected by authors during 2014

that education reduces poverty up to −0.3 %, higher technology score helps in reducing poverty up to 1.9 % and agriculture technology score helps reducing poverty up to 4.5 %. All the other indicators and results are given in Table 5.

4 Econometric Results and Discussion

Although, the unconditional summary statistics and tests in the tables above in general suggest that agriculture related technology adoption may have a positive role in improving household wellbeing, these results are only based on observed mean differences in outcomes of interest and may not be solely due to agriculture related technology adoption. They may instead be due to other factors, such as differences in household characteristics. To measure the impact of adoption; it is necessary to take into account the fact that individuals who adopt agricultural technologies might have achieved a higher level of welfare, even if they had not adopted. As a consequence, we apply propensity score matching methods that control for these observable characteristics to isolate the intrinsic impact of technology adoption on household welfare.

Table 6 provides information about some of the driving forces behind farmers' decisions to adopt agricultural technologies where, the dependent variable takes the value of one if the farmer adopts at least one agricultural related technology and 0 otherwise. The results show that the coefficients of most of the variables hypothesized to influence adoption, have expected signs and they include factors such as the land holding size, gender, education of head of the household, household size, income from agriculture, technology score of household etc. The size of the land owned by the household returned a positive and significant coefficient suggesting that farmers with larger holdings are more likely to adopt than small farmers. According to De Janvry et al. (2001) small farmers will typically prefer new farming systems that are more capital-saving and less risky while large farmers would prefer new farming systems that are more labor saving and they can afford to

Table 6 Determinants of adoption of agricultural related technology

Independent variables	Coefficient	Std. Err.	z
Land holding of household	0.046	0.021	2.190***
Gender of head of household	−0.051	0.022	−2.318***
Age of head of household	−0.021	0.016	−1.313
Education of head of household	0.204	0.099	2.061***
Household size	−0.041	0.017	−2.412***
Income share from agriculture	0.021	0.011	1.909*
Ratio of wet land to total land	1.023	0.711	1.439
Ratio of dry land to total land	1.101	0.928	1.186
Technology score of household	7.981	3.551	2.248***
Participation in MGNREGA	−0.212	0.406	−0.522
Constant	−2.766	1.249	−2.215***
Number of observations	296		
LR $chi^2(10)$	186.840***		
Pseudo R^2	0.482		
Log likelihood	−100.245		

Source Primary data collected by authors during 2014
Note * and *** indicate statistically significant at 10 and 1 % level, respectively

assume risks. In this case small farmers seem to avoid improved varieties due to the high costs associated with the purchasing of improved seed.

Among the explanatory variables, education of head of the household, income from agriculture, higher technology score of households are positively related to the decision to adopt the agriculture related technology. However, gender of head of the household, household size, are negatively related to the decision to adoption of agriculture related technology. Among the other variables, age of the head of the household, ratio of wet land to total land, ratio of dry land to total land and participation in MGNREGA, are not the major determinants of decision to participate in adopting the agriculture related technology at household level. Further, we have conducted the "*balance test*" for balancing of the distribution of relevant covariates between adopters and non-adopters before and after matching. Table 7 presents results of propensity score matching quality indicators before and after matching. The pseudo R^2 also increased significantly from 48 % before matching to about 56 %. This low pseudo R^2, high total bias reduction, and the significant *p*-values of the likelihood ratio test after matching suggest that, the specification of the propensity is successful in terms of balancing the distribution of covariates between the two groups.

Table 8 reports the estimates of the average adoption effects estimated using nearest neighbor matching (NNM) and kernel based matching (KBM) methods. All the analyses were based on implementation of common support and caliper, so that the distributions of adopters and non-adopters were located in the same domain. As suggested by Rosenbaum and Rubin (1985), we used a caliper size of one-quarter

Table 7 Adoption effect on per capita expenditure (results from the PSM)

Matching algorithms	NNMa	NNMb	KBMa	KBMb
Pseudo R^2 before matching	0.482	0.482	0.482	0.482
LR chi^2 before matching	86.840***	86.840***	86.840***	86.840***
Mean standardized bias before matching	21.157	21.157	21.157	21.157
Pseudo R^2 after matching	0.561	0.543	0.541	0.541
LR chi^2 after matching	87.531***	89.541***	88.651***	88.567***
Mean standardized bias after matching	7.969	6.142	4.92	4.884
Total % bias reduction	62.329	71.678	76.797	76.989

Source Primary data collected by authors during 2014
Note *** indicate statistically 1 % level
NNMa Single nearest neighbor matching with replacement, common support, and caliper (0.03); *NNMb* Five nearest neighbors matching with replacement, common support, and caliper (0.03); *KBMa* Kernel based matching with band width 0.03, common support and *KBMb* Kernel based matching with band width 0.06, common support

Table 8 Impact of agricultural related technology adoption

Matching algorithm		Outcome (mean)		ATT
		Adopters	Non-adopters	
aNNM	Per capita expenditure	9.582	9.381	0.200 (2.10)***
	Head count ratio	0.586	0.761	−0.174 (−2.67)***
	Severity of poverty	0.529	0.513	−0.015 (0.10)
bNNM	Per capita expenditure	9.582	9.414	0.167 (2.10)***
	Head count ratio	0.586	0.761	−0.129 (−2.29)***
	Severity of poverty	0.529	0.509	0.020 (0.13)
aKBM	Per capita expenditure	9.582	9.415	0.166 (2.23)***
	Head count ratio	0.586	0.708	−0.121 (−2.20)***
	Severity of poverty	0.529	0.519	0.009 (0.05)
bKBM	Per capita expenditure	9.582	9.415	0.166 (2.29)***
	Head count ratio	0.586	0.709	−0.122 (−2.25)***
	Severity of poverty	0.529	0.523	0.006 (0.03)

Source Primary data collected by authors during 2014
Note *** indicates statistical significance at the 1 %. T-statistics in parenthesis
aNNM single nearest neighbor matching with replacement, common support, and caliper (0.03); *bNNM* Five nearest neighbors matching with replacement, common support, and caliper (0.03); *aKBM* Kernel based matching with band width 0.03, common support and *bKBM* Kernel based matching with band width 0.06, common support, Figures in parentheses at *t*-values

of the standard deviation of the propensity scores. Three outcome variables are used in the analysis such as (1) per capita expenditure, (2) head count ratio, (3) severity of poverty index. The results indicate that, adoption of agriculture related technologies have positive and significant effect on per capita consumption expenditure and negative impact on poverty.

Table 9 Differential impact of adoption by farm size and years of experience

Stratified by farm size (quartiles)	1	2	3	4
Mean impact on household consumption	0.927 (3.0)***	0.273 (1.39)	0.145 (0.86)	0.058 (0.38)
Mean impact on headcount ratio	−0.667 (−4.0)***	−0.318 (−2.32)***	−0.129 (−100)	−0.075 (−0.57)
Mean impact on depth of poverty	−1.066 (−2.77)***	−0.296 (−1.81)*	−0.123 (−0.88)	−0.058 (0.27)
Mean impact on severity of poverty	1.209 (1.58)	0.151 (−0.94)	−0.21 (0.22)	0.061 (−0.13)

Source Primary data collected by authors during 2014
Note * and *** indicate statistically at 10 and 1 % level, respectively

To gain further understanding of the impact of adoption on different groups of adopters, we also examined the differential impacts of adoption by dividing households into quartiles based on consumption, headcount ratio, depth of poverty, and severity of poverty. As observed in Table 9, the impact of adoption on consumption expenditure decreases with farm size. Interestingly, the gain in consumption expenditure and reduction in poverty is highest in the lowest farm-size quartile (1). These findings suggest that adoption of agricultural related technology can contribute to poverty reduction among the near landless households.

5 Conclusion and Policy Implications

The relationship between agricultural technology adoption and welfare is assumed to be straight forward. However, quantifying the causal effect of technology adoption can be quite complex. This paper provides an ex-post assessment of the impact of adoption of agricultural related technology on per capita consumption expenditure and poverty status measured by headcount index in rural India. Our results show that adoption has a positive impact on consumption expenditures and negative on poverty reduction. Though there is a large scope for boosting the role of agricultural technology in anti-poverty policies in rural areas. Implementing poverty alleviation measures, though, is not just the nature of technology but also the inclusion of a poverty dimension into the agricultural research priority-setting. Better targeting of agricultural research on resource-poor producers might be the main vehicle for maximizing direct poverty—alleviation effects. Improved agricultural technology diffusion seems the most effective means of improving agricultural productivity vis-à-vis reducing poverty. Improved rural infrastructure, improved irrigation systems, maintenance of livestock, physical assets, better access to education, secure land tenure, and reasonable access to extension services all play a significant role in encouraging productivity growth and poverty reduction. Technology adoption, however, is constrained by lack of development of market infrastructure, information asymmetry and agriculture extension services. Policies that address these constraints and strengthen local institutions to collectively improve access to technology, credit, and information will increase both the spread and intensity of adoption.

Acknowledgments We would like to thank the participants of the workshop on "*Harnessing Technology for Challenging Inequality*" at Tata Institute of Social Sciences, Mumbai jointly organized with Forum for Global Knowledge Sharing. We gratefully acknowledge Prof. K. Narayanan and Prof. N.S. Siddharthan for comments and suggestions in the earlier draft of this paper. We are grateful to MSSRF-APM Project for the funding support of the sub-project on PDHED at MSE Chennai. We gratefully acknowledge inputs from Prof. U. Sankar, Prof. R.N. Bhattacharyya, Prof. K.R. Shanmugam, and Dr. A. Nambi for the insightful comments and suggestions on the project output. We also grateful acknowledge the respondents for their active participation during primary data collection.

References

Ali, A., & Abdulai, A. (2010). The adoption of genetically modified cotton and poverty reduction in Pakistan. *Journal of Agricultural Economics, 61*(1), 175–192.

Becerril, J., & Abdulai, A. (2010). The impact of improved maize varieties on poverty in Mexico: A propensity score-matching approach. *World Development, 38*(7), 1024–1035.

Bellon, M. R., Adato, M., Becerril, J., & Mindek, D. (2006). Poor farmers' perceived benefits from different types of maize germplasm: The case of creolization in lowland tropical Mexico. *World Development, 34*(1), 113–129.

Binswanger, H. P., & Von Braun, J. (1991). Technological change and commercialization in agriculture: The effect on the poor. *The World Bank Research Observer, 6*(1), 57–80.

David, C. C., & Otsuka, K. (Eds.). (1994). *Modern rice technology and income distribution in Asia*. International Rice Research Institute

De Janvry, A., & Sadoulet, E. (2002). World poverty and the role of agricultural technology: Direct and indirect effects. *Journal of Development Studies, 38*(4), 1–26.

De Janvry, A., Graff, G., Sadoulet, E., & Zilberman, D. (2001). Technological change in agriculture and poverty reduction. *Concept paper for the WDR on Poverty and Development*

Evenson, R., & Gollin, D. (2003). Assessing the impact of the green revolution: 1960 to 2000. *Science, 300*(2), 758–762.

Foster, J., Greer, J., & Thorbecke, E. (1984). A class of decomposable poverty measures. *Econometrica, 52*(3), 761–766.

Imbens, G. W., & Angrist, J. D. (1994). Identification and estimation of local average treatment effects. *Econometrica, 62*, 467–476.

Just, R. E., & Zilberman, D. (1988). The effects of agricultural development policies on income distribution and technological change in agriculture. *Journal of Development Economics, 28* (2), 193–216.

Karanja, D. D., Renkow, M., & Crawford, E. W. (2003). Welfare effects of maize technologies in marginal and high potential regions of Kenya. *Agricultural Economics, 29*(3), 331–341.

Kassie, M., Shiferaw, B., & Muricho, G. (2011). Agricultural technology, crop income, and poverty alleviation in Uganda. *World Development, 39*(10), 1784–1795.
Kijima, Y., Otsuka, K., & Sserunkuuma, D. (2008). Assessing the impact of NERICA on income and poverty in central and western Uganda. *Agricultural Economics, 38*(3), 327–337.
Lee, M. J. (2005). *Micro-econometrics for policy, program and treatment effects*. Advanced Texts in Econometrics, Oxford University Press.
Mendola, M. (2007). Agricultural technology adoption and poverty reduction: A propensity score matching analysis for rural Bangladesh. *Food Policy, 32*(3), 372–393.
Minten, B., & Barrett, C. B. (2008). Agricultural technology, productivity, and poverty in Madagascar. *World Development, 36*(5), 797–822.
Moyo, S., Norton, G. W., Alwang, J., Rhinehart, I., & Demo, M. C. (2007). Peanut research and poverty reduction: Impacts of variety improvement to control peanut viruses in Uganda. *American Journal of Agricultural Economics, 89*(2), 448–460.
Mwabu, G., Mwangi, W., & Nyangito, H. (2006). *Does adoption of improved maize varieties reduce poverty? Evidence from Kenya*. Paper presented at the International Association of Agricultural Economists Conference, Gold Coast, Australia
Rangarajan, C. (2014). *Report of the expert group to review the methodology for measurement of poverty*. Government of India Planning Commission. Available at http://planningcommission.nic.in/reports/genrep/pov_rep0707.pdf
Rosenbaum, P. R., & Rubin, D. B. (1985). Constructing a control group using multivariate matched sampling methods that incorporate the propensity score. *American Statistician, 39*(1), 33–38.
Rubin, D. B. (1978). Bayesian inference for causal effects: The role of randomization. *Annals of Statistics, 6*, 34–58.
Sianesi, B. (2004). An evaluation of the Swedish system of active labour market programmes in the 1990s. *Review of Economics and Statistics, 86*(1), 133–155.
Winters, P., De Janvry, A., Saudolet, E., & Stamoulis, K. (1998). The role of agriculture in economic development: Visible and invisible surplus transfers. *Journal of Development Studies, 345*, 71–97.
Wu, H., Ding, S., Pandey, S., & Tao, D. (2010). Assessing the impact of agricultural technology adoption on farmers' well-being using propensity score matching analysis in Rural China. *Asian Economic Journal, 24*(2), 141–160.

Socio-economy of Mobile Phone Ownership in India

G.D. Bino Paul and Ashutosh Bishnu Murti

1 Introduction

This paper embarks on the theme of growing telecommunication penetration in India. The objective of the paper is to understand the scope and magnitude of the expanding telecommunication, and linkages to advances in development. Quite importantly, this paper plots the linkage between demographic and economic factors and mobile phone penetration rates in India. In fact, the exponential growth and rapid dispersion of mobile telephony has far reaching implications for the economic development for emerging economies such as India. The mobile telephony is providing the access and the connectivity to citizens and transforming their life-style and livelihoods, especially in developing regions. It is important to note that in the context of emerging convergent technologies and systems, mobile phones appear to play the pivotal role in sharing the innovative systems, technologies, knowledge and contents with persons, groups, institutions, and societies, capturing the length and breaths of plural micro-meso-macro scales because, foremost, mobile telephony is a network good that sets context for multiple streams of spill overs and pay-offs. More importantly, inherently, being a network good the utility of mobile phone to a user may be posited to be dependent on the size of the network; higher the size of the network, higher will be utility to the user. Moreover, in comparison with the landline, the mobile set allows the user to embed multiple

G.D. Bino Paul (✉)
School of Management and Labour Studies, Tata Institute of Social Sciences, Mumbai, Maharashtra, India
e-mail: binopaul@gmail.com

A.B. Murti
School of Management and Labour Studies, Tata Institute of Social Sciences, Hyderabad, Telegana State, India
e-mail: murti.ashutosh@gmail.com

N.S. Siddharthan and K. Narayanan (eds.), *Technology*,
India Studies in Business and Economics,
DOI 10.1007/978-981-10-1684-4_9

forms of data simultaneously, thus, deviating from the conventional simplex network to a sort of multiplex arrangement.

According to Rogers (2003), the spread of a new innovation over time typically follows an S-curve, as the early adopters select the technology first, followed by the majority until an innovation is common. Many researchers have estimated the S-curve for mobile phone diffusion using functional forms such as Bass Model (Bass 1969). Existing research also includes the studies focusing on the determinants of mobile phone penetration on a broader scale. Donner (2008) provides an extensive survey of such studies. Rouvinen (2006) studied the mobile phone diffusion across developed and developing countries. Most studies concluded that single standard for mobile platform and competition (number of operators) has positive effects on mobile phone adoption. In a study of 29 countries over a period of 1993–2004, Chakravarty (2007) examined the mobile phone penetration rates in Asia, using panel data analysis. His findings indicate that gross domestic product, per capita income, landline density, number of mobile providers and regulatory policy have positive and statistically significant influence on mobile phone penetration rates.

As argued by Goodman (2005), social capital can be more important concept for the emerging economies compared to the others, as in many cases people in the former have less access to formalised structures of support, and may rely on informal connections instead. Exchange of voice or data using mobile phone can be entirely social, economic, or a combination of the two. Role of social networks and the resulting social capital are neither positive nor negative by nature, but rather a neutral social resource. Quite importantly, Moser (1996) and Narayan (1995) finds that communities having diverse stock of social networks and civic associations are in a stronger position to confront poverty and vulnerability.

Hamilton (2003) explicitly addresses the issue of complementarity or substitution between fixed lines and mobile phones in Africa using a sample of developing countries of Africa. His results suggest that mobile phones are complementary to fixed telephone lines. However, this may be the result of strategic competition within the industry. According to him, "At different stages of cellular development, mobile can play the role of both a substitute for and a complement of main line" (p. 130). Aker and Mbiti (2010) delineates salient aspects of mobile phone coverage in Africa. In the studies on Africa, population density, per capita income and poor quality of landlines seem to have positive correlation with mobile phone coverage. Thus, the review of existing literature shows that a growing body of research has explored a variety of determinants of mobile phone penetration covering various regions and time periods. Most consistently, the factors such as income, fixed lines, industry competition and regulatory policy emerge as the most important predictors of mobile phone penetration. However, the evidence is mixed whether fixed telephone lines are substitutes or complements to mobile phones.

In this paper, we examine the linkage between owning mobile phones and socio-economic status of households in India. While a large chunk of analysis is based on unit records of 66th round of National Sample Survey (NSS)-Consumption Expenditure, we also use the secondary data obtained from Census,

Government of India, 2011 and Telecom Regulatory Authority of India (TRAI). The paper is organised into three sections. Section 2 deals with mobile phone ownership and socio-economic characteristics. Section 3 provides concluding remarks.

2 Mobile Phone Ownership and Socio-economic Characteristics

As shown in Table 1, in 2013, there were 868 Million wireless subscribers in India, reporting a teledensity[1] of 71. While, teledensity in the urban is 141, teledensity in the rural is just 40; there is discernible gap between urban and rural teledensity. Quite important, 88 % of the market is serviced by private operators.

Figure 1 depicts area wise distribution of subscribers in millions. Some of the largest service areas are Utter Pradesh East (75 million), Tamil Nadu (75 million), Maharashtra (71 Million), Andhra Pradesh (67 million), Bihar (61 million), Karnataka (55 million), Gujarat (53 million), Madhya Pradesh (53 million), and so on. Quite interestingly, while in global cities like Mumbai there are 33 million subscribers, the number in the whole North Eastern India is just 9 million.

While the previous pattern depicts area wise subscription of mobile phone services, it is important to figure out proportion of households owning at least one landline or mobile set, not provisioning for multiple subscriptions, to arrive at tele density for every state or union territory in India. We calculate tele density from Census 2011. Among states, Kerala reports highest tele density (0.21), followed by Goa (0.20), Tamil Nadu (0.19), Punjab (0.16), Andhra Pradesh (0.16), Karnataka (0.15), Sikkim (0.15), Mizoram (0.15), Maharashtra (0.15) and so on (Table 2). On the other hand, Chhattisgarh reports the lowest tele density (0.1). States such as Bihar, Uttar Pradesh, Nagaland, Madhya Pradesh, and Jharkhand belong to the cluster of states having noticeably lower tele density. There appears to be a direct relation between tele density and human Development Index (HDI) (Fig. 2).

As shown in Fig. 3, albeit noticeably lower tele density, states like Uttar Pradesh and Bihar appear to form a large chunk of the mobile telephone market; Uttar Pradesh accounts for nearly one sixth of the total market while Bihar's share is 6 %. Other prominent market includes Maharashtra (10 %), Tamil Nadu (8 %), West Bengal (8 %), Karnataka (6 %), Madhya Pradesh (6 %), Gujarat (5 %) and so on.

Table 3 presents proportion of households owning mobile phone in 2009-10, for both the rural and the urban. Among states Kerala reports highest proportion of rural households owning the mobile phone (79 %), Chhattisgarh reports the least (26 %). Quite important, other cases of noticeably lower ownership rates in the rural include Orissa (34 %), Arunachal Pradesh (39 %), Jharkhand (35 %), West

[1]Tele density refers to number of telephone connection for every 100 individuals live with in an area.

Table 1 Indian wireless telecom at Glance

Wireless subscribers	Status
Total wireless subscribers	867.80 Million
Urban subscribers	525.30 Million
Rural subscribers	342.50 Million
Global system for mobile subscribers	794.03 Million
Code division multiple access subscribers	73.78 Million
Market share of private operators	87.76 %
Market share of public sector unit operators	12.24 %
Tele density	70.85
Urban tele density	140.67
Rural tele density	40.23

Source Telecom Regulatory Authority of India (Data as on 31st March, 2013)

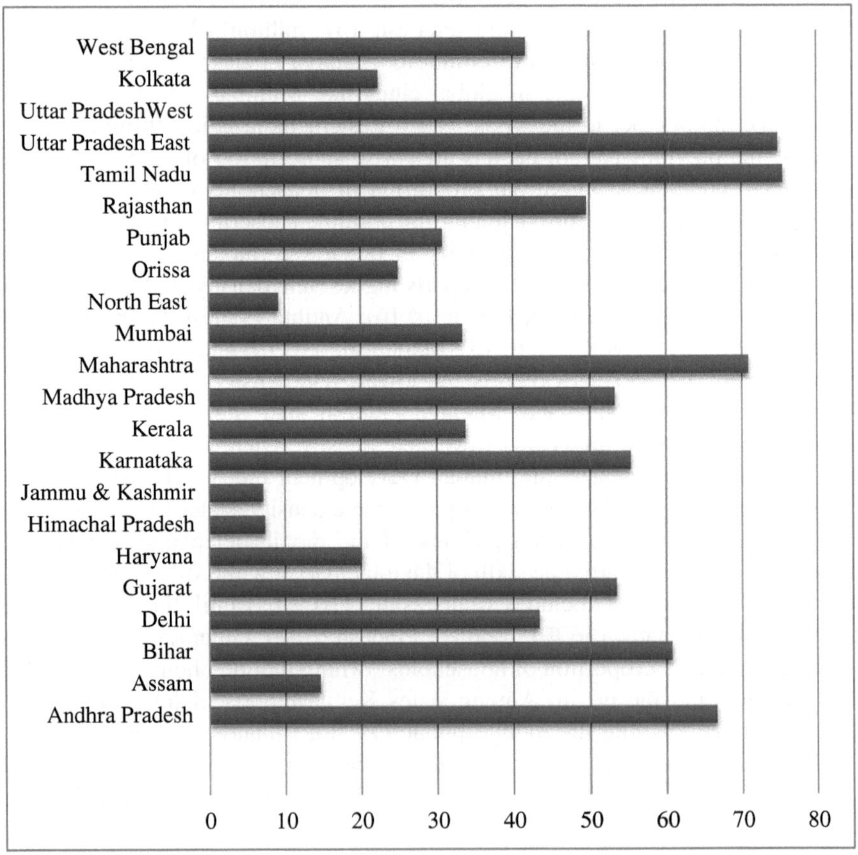

Fig. 1 Area wise distribution of subscribers in millions. *Source* Telecom Regulatory Authority of India (Data as on 31st March, 2013)

Table 2 Teledensity and human development index (HDI)

State	Tele density	HDI 2006	State	Tele density	HDI 2006
Andaman and Nicobar Islands	0.205	0.708	Lakshadweep	0.155	0.697
Andhra Pradesh	0.157	0.585	Madhya Pradesh	0.095	0.529
Arunachal Pradesh	0.091	0.647	Maharashtra	0.146	0.689
Assam	0.098	0.595	Manipur	0.113	0.702
Bihar	0.101	0.507	Meghalaya	0.078	0.629
Chandigarh	0.199	0.784	Mizoram	0.147	0.688
Chhattisgarh	0.068	0.549	Nagaland	0.107	0.700
Dadra and Nagar Haveli	0.134	0.677	Orissa	0.092	0.537
Daman and Diu	0.214	0.700	Puducherry	0.195	0.725
Delhi	0.181	0.740	Punjab	0.160	0.668
Goa	0.197	0.764	Rajasthan	0.130	0.541
Gujarat	0.139	0.634	Sikkim	0.153	0.665
Haryana	0.148	0.643	Tamil Nadu	0.192	0.666
Himachal Pradesh	0.177	0.667	Tripura	0.110	0.663
Jammu and Kashmir	0.112	0.590	Uttar Pradesh	0.110	0.528
Jharkhand	0.090	0.574	Uttarakhand	0.148	0.652
Karnataka	0.154	0.622	West Bengal	0.108	0.642
Kerala	0.207	0.764	Total	0.130	0.614

Source HDI from UNDP and teledensity from census 2011

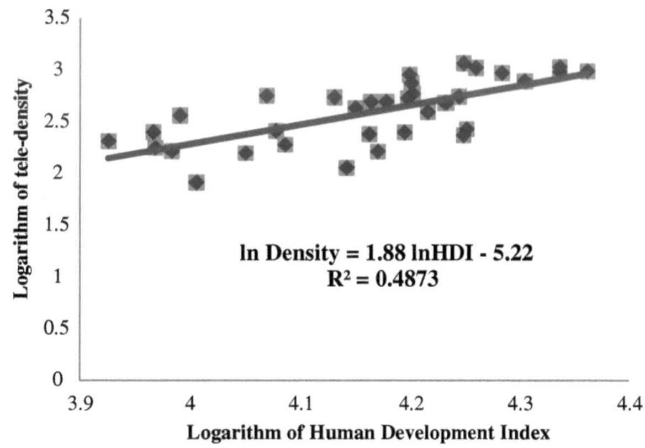

Fig. 2 Teledensity and human development index (HDI). *Source* Estimated from the author using UNDP and census 2011 data

Bengal (40 %), Bihar (48 %), and Meghalaya (49 %). Overall, in India share of rural households who own mobile phone is 55 %, while 81 % of urban households in India own the mobile set.

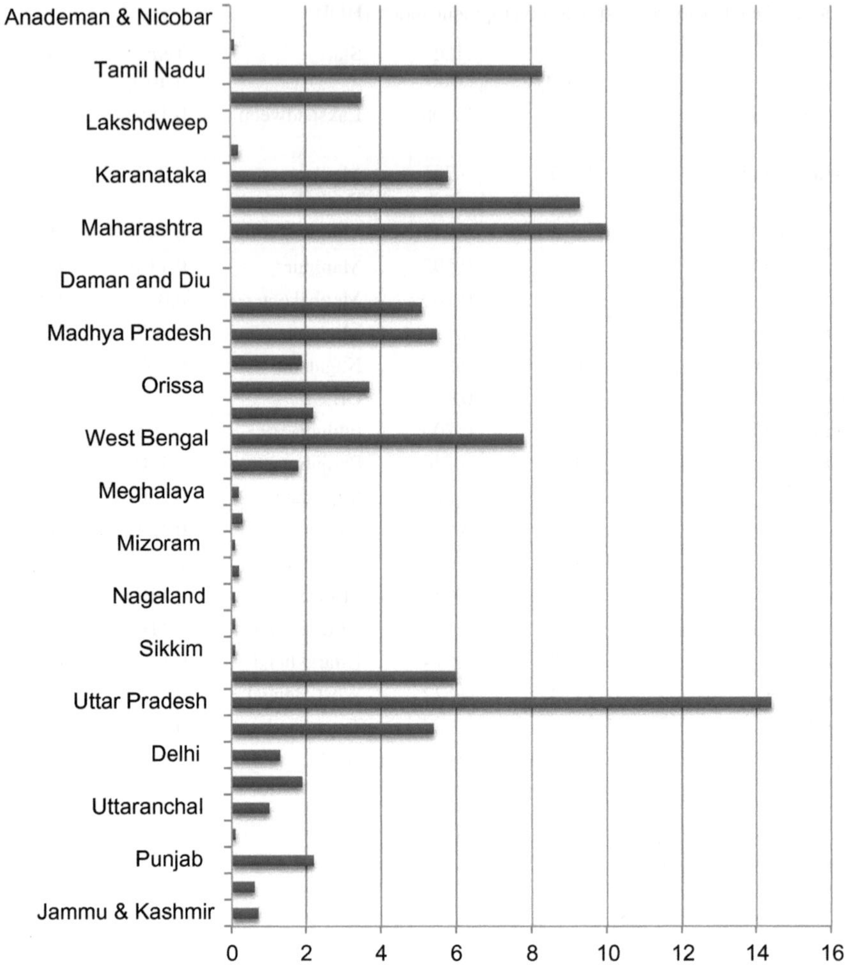

Fig. 3 Distribution of mobile ownership. *Source* Estimated from the unit records, NSS 66th round on consumption expenditure (2011), Number of observation = 94,311

Table 4 cross-tabulates proportions of households by age interval. We categorise head of household's age to three categories: 34 and below, 35–59 and 60 and above. Overall, in India a two third of households are in the age group 35–59 own the mobile phone, while ownership rates for 34 and below and 60 and above are 61, 59 %, respectively. Interestingly, this pattern is not valid in states like Tamil Nadu where ownership rate falls with age; while ownership rate in respect of 34 and below is 77, ownership rates for 35–59 and 60 and above are 76.5 and 50.5 %, respectively. On the contrary, in West Bengal, there appears to be a direct relation between ownership percentage and age.

Table 3 Proportion of households owning mobile phone

State	Rural (%)	Urban (%)	State	Rural (%)	Urban (%)
Jammu and Kashmir	74.6	83.3	West Bengal	40.4	76.3
Himachal Pradesh	78.2	88.0	Jharkhand	34.9	73.7
Punjab	75.8	80.8	Orissa	33.9	71.9
Chandigarh	91.5	93.1	Chhattisgarh	25.8	72.9
Uttaranchal	75.0	87.2	Madhya Pradesh	44.2	79.6
Haryana	76.2	85.1	Gujarat	64.4	85.1
Delhi	78.4	88.5	Daman and Diu	76.9	95.0
Rajasthan	71.5	85.3	Dadra and Nagar Haveli	60.4	97.2
Uttar Pradesh	55.4	78.4	Maharashtra	52.6	83.6
Bihar	47.7	75.6	Andhra Pradesh	55.1	80.2
Sikkim	66.7	94.6	Karnataka	58.4	84.1
Arunachal Pradesh	38.6	74.5	Goa	68.0	86.6
Nagaland	68.4	90.0	Lakshadweep	79.4	88.9
Manipur	54.9	74.8	Kerala	78.5	85.2
Mizoram	58.7	90.7	Tamil Nadu	64.7	79.4
Tripura	55.0	85.4	Pondicherry	70.5	84.6
Meghalaya	48.6	82.3	Andaman and Nicobar	78.0	94.0
Assam	49.9	85.0	India	54.8	81.4

Source Estimated from the unit records, NSS 66th round on consumption expenditure (2011), Number of observation = 94,311

In Table 5 the proportion of households owning mobile phone is computed with respect to social category; there are four social categories-Scheduled Tribes (ST), Scheduled Castes (SC), Other Backward Classes (OBC), and Others. In 2009–2010, in India, while the proportion of ST households owning the mobile phone is the least (42 %) compared to other social categories the category others reports the highest category (75 %). It appears most of the major states in terms of the size of the mobile ownership show more or less same pattern. However, in some union territories the above-mentioned pattern appears to be not valid; rather a topsy-turvy distribution is reported. States such as Chhattisgarh (18 %), Orissa (20 %) and Jharkhand (29 %) report discernibly lower ownership rates for ST.

Table 6 provides probit estimates of mobile phone ownership. Here, we specify mobile phone ownership as a function of place of residence, social category, religion, having Internet connection, educational attainment, age, state, constant and error. Quite important, there is a statistically significant positive coefficient in favour of the urban over the rural in owning the mobile phone. As regards the social category, all groups report statistically significant positive coefficients over the reference group "Scheduled Tribe". In the case of religion, except Buddhist and others, all other groups report statistically significant positive coefficient over

Table 4 Proportion of households owning mobile phone (Rural + Urban) by age interval

State	34 and below (%)	35–59 (%)	60 and above (%)	State	15–34 (%)	35–59 (%)	60 and above (%)
Jammu and Kashmir	72.6	77.8	76.4	West Bengal	43.4	50.8	58.2
Himachal Pradesh	86.4	80.0	73.7	Jharkhand	41.7	42.6	46.0
Punjab	65.3	83.9	71.5	Orissa	33.7	40.3	47.8
Chandigarh	89.6	94.1	98.4	Chhattisgarh	30.2	36.2	41.8
Uttaranchal	85.6	76.4	68.3	Madhya Pradesh	50.8	55.3	49.3
Haryana	74.2	81.3	77.4	Gujarat	68.8	76.4	65.4
Delhi	84.1	91.6	90.7	Daman and Diu	88.0	86.9	69.4
Rajasthan	68.8	78.9	71.1	Dadra and Nagar Haveli	64.6	75.2	95.3
Uttar Pradesh	58.0	62.5	57.5	Maharashtra	64.8	70.9	57.4
Bihar	50.5	50.6	53.6	Andhra Pradesh	64.0	67.2	49.8
Sikkim	68.1	74.0	64.9	Karnataka	72.2	67.8	71.0
Arunachal Pradesh	49.9	49.5	22.2	Goa	41.3	81.1	77.3
Nagaland	78.3	78.6	50.5	Lakshadweep	98.7	84.0	84.3
Manipur	48.2	59.7	69.6	Kerala	80.3	84.4	72.5
Mizoram	58.7	78.7	75.1	Tamil Nadu	77.3	76.5	50.5
Tripura	51.6	63.0	66.9	Pondicherry	69.5	88.8	69.1
Meghalaya	46.7	54.9	69.4	Andaman and Nicobar	84.6	84.3	87.0
Assam	44.7	55.5	61.2	India	60.7	65.2	59.4

Source Estimated from the unit records, NSS 66th round on consumption expenditure (2011) Number of observation = 94,311

Hindu. Those households not having Internet connection report a statistically significant negative coefficient over the households having Internet connection. As far as educational attainment is concerned, all categories report statistically significant positive coefficient over the reference category "not literates". Over the reference category age group 34 years and below other age group report statistically significant positive coefficients. What the results signify is that ownership of mobile phone appears to be mired in socio-economic-demographic characteristic such as place of residence, social category, religion, having Internet connection, educational attainment, and age.

Table 5 Proportion of households owning mobile phone (Rural + Urban) by social category

State	Scheduled Tribes (%)	Scheduled castes (%)	Other backward classes (%)	Others (%)
Jammu and Kashmir	63.7	64.4	75.7	79.0
Himachal Pradesh	76.3	71.6	75.7	83.9
Punjab	62.7	65.1	77.4	88.9
Chandigarh	100.0	84.0	88.2	96.1
Uttaranchal	62.8	54.6	75.7	84.3
Haryana	68.8	60.6	79.4	91.2
Delhi	45.8	86.4	79.4	92.6
Rajasthan	69.8	65.8	76.3	85.4
Uttar Pradesh	71.7	46.6	60.0	76.2
Bihar	49.3	35.7	49.6	71.9
Sikkim	67.2	77.9	69.3	95.6
Arunachal Pradesh	39.3	47.2	83.3	60.5
Nagaland	73.6	100.0	75.1	92.2
Manipur	42.3	51.3	69.5	74.4
Mizoram	74.5	69.8	91.8	65.0
Tripura	48.3	55.0	67.8	73.6
Meghalaya	53.8	88.9	47.3	65.1
Assam	40.3	54.4	54.5	59.8
West Bengal	35.5	39.1	53.1	56.0
Jharkhand	28.7	31.8	50.2	64.9
Orissa	19.8	29.9	43.5	61.0
Chhattisgarh	18.1	34.9	39.4	57.9
Madhya Pradesh	28.8	45.8	55.5	79.2
Gujarat	48.7	64.4	71.2	88.6
Daman and Diu	36.4	100.0	99.2	82.2
Dadra and Nagar Haveli	59.9	70.8	100.0	99.5
Maharashtra	45.4	57.1	61.8	77.2
Andhra Pradesh	38.0	57.3	61.2	72.5
Karnataka	58.8	52.5	71.6	75.0
Goa	36.4	39.1	92.9	83.4
Lakshadweep	84.2	83.6	62.9	100.0
Kerala	64.7	71.8	81.6	80.9
Tamil Nadu	66.7	60.2	73.4	86.7
Pondicherry	100.0	63.1	83.5	81.1
Andaman and Nicobar	65.1	–	88.2	85.4
India	41.6	51.3	64.2	74.5

Source Estimated from the unit records, NSS 66th round on consumption expenditure (2011), Number of observation = 94,311

Table 6 Determinants of mobile ownership

Probit regression				
Dependent variable: mobile handset ownership	Coefficient	Robust Standard Error	z	Level of significance (probability of type 1 error)
Living in Urban sector	0.206	0.010	20.63	1 %
Social category (Reference category = Scheduled Tribe)				
Scheduled castes	0.102	0.018	5.66	1 %
Other backward classes	0.313	0.016	18.99	1 %
Others	0.428	0.017	24.46	1 %
Religion (Reference category = Hindu)				
Muslim	0.049	0.014	3.34	1 %
Christian	0.142	0.021	6.6	1 %
Sikh	0.600	0.037	15.86	1 %
Janis	0.380	0.134	2.82	1 %
Buddhist	−0.010	0.044	−0.25	Not significant
Others	−0.374	0.052	−7.11	1 %
Not having internet connection	−0.592	0.055	−10.67	1 %
Educational attainment (Reference category = Not Literate)				
Just literate	0.291	0.015	18.38	1 %
Primary	0.423	0.014	28.99	1 %
Middle	0.671	0.014	46.44	1 %
Secondary	0.978	0.016	59.63	1 %
Higher secondary/diploma	1.202	0.020	59.09	1 %
Graduate	1.483	0.025	58.97	1 %
Post graduate	1.603	0.046	34.55	1 %
Age (Reference category = 15–34)				
35–59	2.448	0.171	14.27	1 %
60 and above	2.437	0.171	14.19	1 %
State dummy (Yes)				
Constant	−2.146	0.182	−11.79	1 %

Estimated from national sample survey 66th round unit records on consumption expenditure (2011), Number of observation = 94,311, Wald chi2 (22) = 12873.85, Prob > chi2 = 0, Pseudo R2 = 0.1438, Log pseudo likelihood = − 48136.481

Moreover, incremental consumption expenditure by mobile owning households over the not owning ones, as plotted in Fig. 4, appears to be positive for all states/union territories. Among the states, Sikkim reports highest positive differential (87 %) in favour of households who own mobile phone over those who do not own mobile phone, Punjab and Karnataka report the least (25 %).

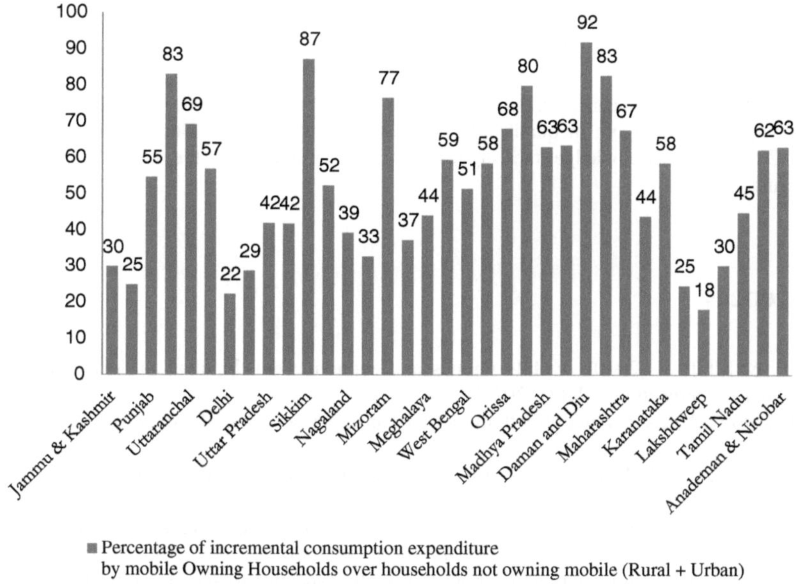

Fig. 4 Percentage of incremental consumption expenditure by mobile owning households not owning (Rural + Urban). *Source* Estimated from the unit records, NSS 66th round on consumption expenditure (2011), Number of observation = 94,311

3 Concluding Remarks

Drawing cues from patterns and inferences that have emanated from the analysis of data, quite vividly, there appears to be a direct linkage between the ownership of mobile phones and socio-economic status, in particular the economic status. Although the positive linkage that we argue may have been impacted by the factor of serendipity there appears to be a systematic pattern that elucidates the pivotal aspects in the socio-economy that determines the ownership of technology embedded artefacts such as mobile phone. However, particularly in the context of fast growing economic geographies such as India, patterns point to discernible gaps or inequalities in owning technology embedded goods such as mobile phones that stem from diverse socio-economic-demographic characteristics such as place of residence, social category, and religion, having Internet connection, educational attainment, age, and the state. Unequivocally, as the plot of human development index against tele density points to the usefulness of network goods such as mobile phones to let households as well as persons progressing to higher order socio-economic strata. Presumably, in order to reap the benefit of the direct relation between owning the mobile set and the socio-economic status entails more creative strategies that make access to network goods more inclusive, particularly in the context of the emerging digital convergent technologies and systems.

Acknowledgments The earlier version of this paper was presented at the workshop 'Harnessing Technology for Challenging Inequality', held at Tata Institute of Social Sciences, Mumbai on 7th March 2015. We acknowledge the suggestions and comments of the Chair/Discussant and participants.

References

Aker, J., & Mbiti, I. (2010). Mobile phones and economic development in Africa. *Journal of Economic Perspectives., 24*(3), 207–232.

Bass, F. (1969). A new product growth model for consumer durables. *Management Science, 15*(1), 215–227.

Chakravarty, S. (2007). Determinants of mobile penetration in Asia-a panel study. *Indian Journal of Economics and Business., 6*(2), 147–165.

Donner, J. (2008). Research approaches to mobile use in the developing world: A review of the literature. *The Information Society: An International Journal., 24*(3), 140–159.

Goodman, J. (2005). Linking mobile phone ownership and use to social capital in rural South Africa and Tanzania. Africa: The impact of mobile phones. Vodafone Policy Paper Series 2 (pp. 53–65), March 2005. Vodafone Group.

Hamilton, J. (2003). Are mainlines and mobile phone lines substitutes or complements? Evidence from Africa. *Telecommunications Policy, 27*(1–2), 109–133.

Moser, C. (1996). Confronting crisis: A comparative study of household responses to poverty and vulnerability in four poor Urban communities. Environmentally Sustainable Development Studies and Monographs Series 8. Washington, D.C.: World Bank.

Narayan, D. (1995). *Designing community-based development*. Social Development Paper 7. World Bank, Environmentally and Socially Sustainable Development Network, Washington, D.C.

Rogers, E. M. (2003). *Diffusion of innovations*. New York: Free Press.

Rouvinen, P. (2006). Diffusion of digital mobile telephony: Are developing countries different? *Telecommunications Policy, 30*, 46–63.

ICTs as Enabler in Higher Education

Kaushalesh Lal and Shampa Paul

1 Introduction

The benefits of use of Information and Communication Technologies (ICT) in any sphere of life are no longer unknown. That is the reason why ICT revolution has diffused so rapidly in India and elsewhere. However, the uneven diffusion of ICT created digital divide among different segments of the society. For instance, there is digital divide between rich and poor (Lal and Paul 2004) and there exists rural-urban digital divide. The field of education has not been untouched with this divide. Institutions located in urban areas have better access to new technologies compared to the institutions located in rural or semi-urban areas. Hence urban academic institutions can equip students with better knowledge of ICT.

Before proceeding further, few empirical studies done on the use of ICT in higher education are discussed. One such study by Brown and Liedholm (2002) finds that ICT in higher education can be more useful in some activities such as student-university communication, searching of reading material, online accessibility of literature, and use of ICT led teaching methods. The authors did not find much use of very advanced ICT tools such as virtual classrooms. On the other hand, several studies found challenges as well as opportunities in using ICT in higher education (James and Hopkinson 2009; Loing 2005). James and Hopkinson (2009)

K. Lal (✉)
UNU-MERIT, Maastricht, The Netherlands
e-mail: klal_ieg@yahoo.com

K. Lal
Institute for Human Development, New Delhi, India

S. Paul
Society for Development Studies, India Habitat Center, New Delhi, India
e-mail: shampa.paul147@gmail.com

N.S. Siddharthan and K. Narayanan (eds.), *Technology*,
India Studies in Business and Economics,
DOI 10.1007/978-981-10-1684-4_10

found that use of ICT in higher education has growing social impact. On the other hand, Loing (2005) argued one of the several challenges as “the diversification of learner population, and of institutions and programmes; the permanent upgrading of knowledge contents and a growing need for lifelong learning: here again universities have to meet this challenge and can probably do it only with the help of ICT educational applications.” Lal (2007) in his book emphasized that more ICT savvy managers and workers are needed in the era of globalization.

In Indian context, it was in early 1990s, the Department of Electronics set up a consortium of top Indian institutions such as Indian Institute of Technologies and Indian Institute of Science, Bangalore. An educational network called Education and Research NETwork (ERNET) came into existence due to efforts of these institutions and they were the first to use digital technologies. Realizing the tremendous potential of ICT, the Internet was formally launched in India on August 15, 1995 for public and other institutions. Since then ICT have penetrated in almost all the academic institutions located particularly in urban areas. The penetration of ICT in academic institutions located in rural areas is still dismal despite the several initiatives taken by public and private sector. One of the leading private sector companies that are engaged in diffusion of ICT in education is EDUCOMP. Another path breaking initiative undertaken by TATAs is TATA ClassEdge. They have developed very cost effective ICT led teaching tools such as touch screen boards and Internet and Intranet based teaching methods. The Government of India has also taken up several initiatives so that every student can reap the benefits of ICT. The launch of special satellite, EDUSAT, exclusively for education is a case in point.

This paper seeks to identify major impediments that have resulted in slow penetration of ICT in higher educational institutions in rural India and also to identify mechanism so that rural-urban digital divide is reduced if not eliminated. Before proceeding further, some background of Sitapur district which has been selected for the study is presented. Geographical location of the district on the map of Uttar Pradesh is depicted in Fig. 1.

As far as mapping of the district from higher education institutions point of view is concerned, there are fifty-four[1] degree colleges in the district. Out of these, two are government degree colleges and four are aided by the government. The remaining are self-financed institutions constituting 88.89 % of the total colleges. The strength of students is likely to follow the similar scenario. Almost all self-financed degree colleges are located in rural areas. Hence any initiative taken by the university or government to reduce rural-urban digital divide would need to strengthen self-financed institutions as well. Main findings of the study are highlighted in the following Box I.

---

[1]The information has been collected from the website of Chhatrapati Sahu Ji Maharaj University, Kanpur. It may be mentioned that there is only one engineering college in the district which is affiliated to Gautam Buddha Technical University and rest all are affiliated to CSJMU, Kanpur.

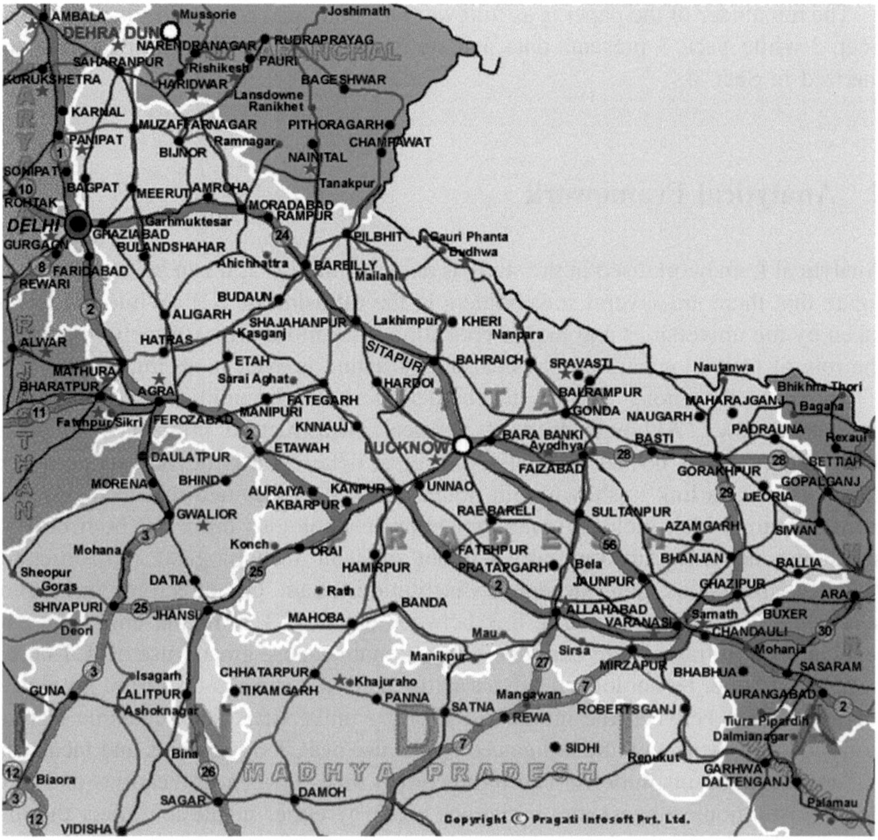

Fig. 1 Location of Sitapur

Box I: Main Findings of the Study

- Initiatives Taken by the University for Digital Student-University Interface are Appreciated by Majority of Students
- There is Need for Regular Faculty Upgradation Programmes to Motivate Them to use ICT led Teaching Methods
- Awareness Among Students is also Needed so That They can Appreciate and Make use of the ICT in a More Effective Manner
- Cyber-café on the Basis of Public-Private-Partnership Model may be Set Up in the Premises of Institutions
- Having a Compulsory Paper about Digital Technologies at the Graduation Level on the Lines of Environmental Paper is Expected to Help Students to Acquire More Knowledge

The remainder of the paper is as follows. The analytical framework is depicted in Sect. 2 while Sect. 3 presents data and statistical analysis. The findings are summarized in Sect. 4.

2 Analytical Framework

Analytical framework used in the study is depicted in Fig. 2. It can be seen from the figure that there are several stakeholders in the diffusion of ICT. While initiatives taken by the universities and higher education institutions are extremely important, the role of technological and physical infrastructure is even more important.

The causality among various stakeholders is shown by unidirectional and bidirectional arrows. Although lack of physical infrastructure influences all other stakeholders, it has not been discussed in great detail as it is beyond scope of the paper. Hence its link is shown with dotted lines. It is assumed that technological infrastructure which includes communication technologies influence both universities and academic institutions. Without efficient technological infrastructure, neither universities nor academic institutions can be effective in digital student-university interface. The link between use of ICT in higher education and technological infrastructure could be bidirectional because greater use of ICT could necessitate better technological infrastructure and vice versa is also true. Similarly, the association between the initiatives taken by universities and academic institutions could influence and be influenced by the use of ICT by students and faculty in higher education institutions. The bidirectional link between universities and academic institutions suggest that initiatives taken by either institution affect the initiatives taken by the other.

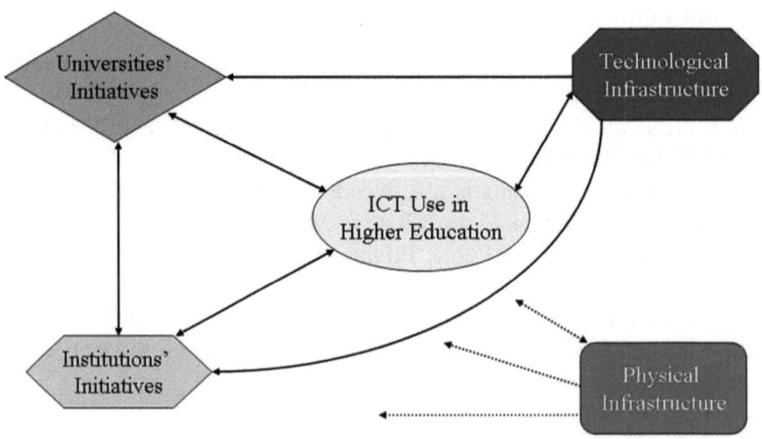

Fig. 2 Analytical framework

The data are analysed within the framework depicted in Fig. 2. However, due to lack of resources it has not been possible to include information from technological infrastructure providing agencies in the analysis. The sources of data and statistical analysis are presented in the following section.

3 Data and Statistical Analysis

The primary data for the study has been collected from four higher education institutions located in Sitapur district of Uttar Pradesh which has been selected as one of the six e-districts by the Government of Uttar Pradesh. These institutions are: Shri D.P. Verma Memorial Degree College, RMP Degree College, Sitapur Sikshan Sansthan, and Institute of Engineering and Technology. Shri D.P. Verma Memorial Degree College falls in rural area of the district while others are located in urban areas. All other institutions, except RMP Degree College, are self-financed institutions. Except engineering college which is affiliated to Gautam Buddha Technical University, Uttar Pradesh, all other institutions are affiliated to Chhatrapati Sahu Ji Maharaj University, Kanpur.

A set of 120 questionnaires were distributed to the students studying in various courses in these institutions. Out of the total, 72 completed questionnaires were received resulting into the response rate of roughly 58 %. The survey was conducted during October 2012 and January 2013. Before discussing statistical results and findings of the study, the profile of respondents is presented in Table 1.

Table 1 Profile of respondents

Academic information	Institution/course	Residence type				Total	Chi-Square
		Rural		Urban			
		No.	%	No.	%	No.	
Institutional affiliation							0.00**
	SSS	2	28.6	5	71.4	7	
	IET	–	–	4	100.0	4	
	RMP	34	75.6	11	24.4	45	
	SDPVMDC	4	25.0	12	75.0	16	
Course studying							0.061*
	BA	18	58.1	13	41.9	31	
	BBA	–	–	5	100.0	5	
	BSc	10	71.4	4	28.6	14	
	MA	9	69.2	4	30.8	13	
	MEd	2	33.3	4	66.7	6	
	MSc.	1	33.1	2	66.7	3	
Total		40	55.6	32	44.4	72	

Note % Row percentage

**1 % level of significance; *10 % level of significance

SSS Sitapur Sikshan Sansthan; *IET* Institute of Engineering and Technology; *RMP* RMP Degree College; *SDPVMDC* Shri D.P. Verma Memorial Degree College

It can be seen from the table that largest percentage (62.5 %) of respondents comes from RMP Degree College followed by 22.2 % from Shri D.P. Verma Memorial Degree College. The minimum participation (5.56 %) is from engineering college. In terms of courses in which students were studying, largest percentage (43.06 %) comes from students who opted for Bachelor of Arts (BA) courses followed by Bachelor of Science (BSc) and Master of Arts (MA) courses 19.44 and 18.06 % respectively. The lowest percentage (4.17 %) of respondents was from Master of Science (MSc) programme. The table also shows that 55.6 % of students had rural background while 44.4 % belonged to urban areas. The association between institutional affiliation and background of students is highly statistically significant. The Chi-square statistics is significant at 1 % level suggesting that institution that is located in rural area had the large percentage of students with rural background. The association of background of students with courses opted is also significant at 10 % level. Hence it can be concluded that more students belonging to rural areas were doing non-technical courses such as BA and BSc while urban students preferred technical courses.

Table 2 presents the availability of technological infrastructure encompassing network technologies, Internet speed, and type of communication technologies.

Internet and Local Area Network (LAN) are the only network technologies considered in the study as Wide Area Network (WAN) was not available in Sitapur district. The results presented in Table 2 suggest that availability of network

Table 2 ICT infrastructure in institutions

ICT infrastructure		Residence type				Total		Chi-Square
Components	Type/speed	Rural		Urban				
		No.	%	No.	%	No.	%	
Networks								
	Internet	29	72.5	27	84.4	56	77.8	0.228
	LAN	2	5.0	2	6.3	4	5.6	0.818
Internet speed								0.008**
	≤100 M	21	52.5	8	25.0	29	40.3	
	101–1000 M	2	5.0	9	28.1	11	15.3	
	>1 G							
Mode of communication								
	Telephone	20	50.0	8	25.0	28	38.9	0.031*
	Dedicated line	2	5.0	1	3.1	3	4.2	0.692
	Wireless	4	10.0	18	56.3	22	3.6	0.000**
	VSAT[a]			3	9.4	3	4.2	0.048*

Note % Percentage of respondents to the total in each category whose response of affirmative; *M* Million bits per second; *G* Giga bits per second

**Significant at 1 % level; *Significant at 5 % level

[a]VSAT (Very small aperture terminal) is a satellite communication system

technologies were not associated with the background of the students. This may be attributed to the fact that network technologies are accessible only in urban areas and students, irrespective of their background, had to use them in the district headquarters. However, Internet speed is highly statistically significant with the background of students. The cost of Internet use may be attributed to this phenomenon. The students from rural background might have preferred slow speed of Internet because of cost considerations. It may be mentioned here that high speed Internet is costlier.

Table 2 also presents interesting findings related to mode of communication for the Internet. More rural students preferred telephone based Internet connectivity while students with urban background preferred wireless connectivity. The results suggest that association between background of students and mode of communication is highly significant.

Enabling environment and availability of digitized information through Internet or Intranet are expected to encourage students to use more modern technologies. In this context, the data were collected from students about provision of digital communication with the affiliating Universities. Table 3 presents the awareness of students regarding digitized facilities provided by their Universities.

It can be seen from Table 3 that large percentage of students is aware that examination form, enrolment form, digital library, and results are available on the website of respective universities. It may be worth mentioning here that Student Login facility has been provided by Chhatrapati Shahu Ji Maharaj University from the academic year of 2010–2011. Hence it is not surprising that merely 36.1 % of students were aware of its existence on the website of the University. This facility is a major improvement in communication between student and the university. Although common e-mail communication with the University has been available to students for the last several years, Student Login provides individual communication with the University authorities. The login and password would remain valid for a student as long as he/she is student of the University irrespective of the course being persuaded.

Mere communication with the University is not the only activity where ICT can be useful. There are stand-alone activities where new technology can be used for improved learning. Here comes the use of ICT by faculty in higher education

Table 3 Digital communication provision of the university

Digitized activities	No.	Percentage
Student login	26	36.1
Examination form	51	70.8
Enrolment form	29	40.3
Results	52	72.2
Reading material	5	6.9
Lectures	3	4.2
Digital library	53	73.6
Provision for communication with the university	5	6.9

Table 4 ICT use by the faculty for teaching

Activities	No.	Percentage
Preparation of class notes	24	33.3
Preparation of literature	14	19.4
For better illustration of ideas	28	38.9
For better explanation of concepts	26	36.1

institutions. Table 4 presents the extent of ICT use by the faculty members of the sample institutions.

The results depicted in Table 4 present a very dismal picture in the context of use of ICT by faculty. Roughly one third of students feel that ICT were being used by faculty for the teaching activities such as, preparation of class notes, better illustration of ideas, and better explanation of concepts. Merely 19.4 % of students feel that ICT were being used for preparation of literature related to their studies. In subsequent section the reason for lesser use of ICT by the faculty are explored. The results related to non-use of ICT are presented in Table 5.

The results presented in Table 5 suggest that lack of ICT infrastructure is considered the most important impediment in the use of ICT by faculty in these institutions. The second most important impediment is the lack of motivation of faculty. This is really an important finding of the study because it substantiates the argument of several scholars who argue that faculty upgradation programmes are very important for improving the teaching standards in the higher education. The lack of awareness of ICT led teaching tools and lack of appreciation by students are other two hindrances in the adoption of new teaching aids. Hence it can be inferred that the awareness programme related to ICT aided teaching might be useful for the students as well as faculty.

Teaching is not the only activity where ICT can benefit students. There are several non-teaching activities where it can be used by students. Table 6 presents the findings related to purpose of ICT use by students.

Table 6 presents not only purpose of ICT use but also its association with the background of students. It is found that most common purpose of using Internet by the students cutting across the background is to see their course results. The second most important activity is to download or online submission of examination forms. There is huge variation in percentage of students who use ICT for filling examination forms in relation to their social background. Thirty percent of rural background students used ICT while more than double the percentage (62.5 %) of urban background students preferred use of Internet in filling examination forms. It may

Table 5 Reasons for not using ICT by the faculty

Activities	No.	Percentage
Lack of ICT infrastructure in the institution	22	30.6
Teachers are not aware of ICT led teaching tools	13	18.1
Lack of motivation	16	22.2
Lack of appreciation by students	11	15.3

Table 6 Purpose of ICT use by students

Purposes	Rural		Urban		Total		Chi-Square
	No.	%	No.	%	No.	%	
Downloading examination form	12	30.0	20	62.5	32	44.4	0.006**
Downloading enrolment form	5	12.5	9	28.1	14	19.4	0.096*
To get teachers' note on Pen drive/CD	–	–	8	25.0	8	11.1	0.001**
Results	23	57.5	19	59.4	42	58.3	0.873
Reading material	5	12.5	13	40.6	18	25.0	0.006**
Offline lectures	2	5.0	3	9.4	5	6.9	0.468
Communication with the university	2	5.0	12	37.5	14	19.4	0.001**

Note **1 % level of significance; *10 % level of significance

be worth mentioning that usually first year examination forms are filled manually and subsequent years it is online submission in CSJMU. The association between filling of examination form and background of students is statistically very significant.

The third important activity for which ICT was being used by students is downloading reading material through Internet. The relationship between downloading of reading material through Internet and background of students is found to be highly significant (1 % level of significance). Higher percent of urban background students (40.6 %) were using ICT for downloading reading material while the percentage of rural background students was very low (12.5 %). Same percentage (19.4 %) of total students irrespective of their background used ICT for downloading enrolment form and communicating with the University. It can be seen from Table 6 that the percentage of urban background students that used ICT in these activities was higher than those who had rural background. The relationship is statistically significant. The results presented in Table 6 suggest that in general students with rural background are less prone to usage of ICT compared to students who had urban background. Easy accessibility of ICT and lack of awareness and knowledge of ICT to rural background students could be attributed to this phenomenon.

Having analysed the usage pattern of ICT by students, their opinion on the level of satisfaction is examined. The findings are presented in Table 7.

The results presented in Table 7 show that fairly large percentage of students (62.5 % of rural and 65.6 % of urban background) were satisfied with digitized facilities provided by the universities. On the other hand, 44.4 % of students were not satisfied with ICT infrastructure available in the institutions. Among non-satisfied students, larger percentage (55.0 %), come from rural background. The association between level of satisfaction related to ICT infrastructure and background of students is statistically significant.

Table 7 Level of satisfaction of ICT use

Activities	Satisfaction level	Rural		Urban		Total		Chi-Square
		No.	%	No.	%	No.	%	
Are you satisfied with the digitized facilities provided by the university on its website?								
	S	25	62.5	21	65.6	46	63.9	0.052*
	NS	5	12.5	1	3.1	6	8.3	
	NE	1	2.5	6	18.8	7	9.7	
Are you satisfied with ICT facilities provided by the institution?								
	S	7	17.5	13	40.6	20	27.8	0.057*
	NS	22	55.0	10	31.3	32	44.4	
	NE	4	10.0	6	18.8	10	13.9	
Are you satisfied with the extent of ICT use by faculty?								
	S	15	37.5	10	31.3	25	34.7	0.570
	NS	14	35.0	16	50.0	30	41.7	
	NE	4	10.0	3	9.4	7	9.7	
Are you satisfied with ICT infrastructure?								
	S	5	11.3	14	42.3	19	25.2	0.018**
	NS	9	22.5	30	93.8	39	56.3	
	NE	2	5.0	7	21.9	9	12.6	

S Satisfied; *NS* Not Satisfied; *NE* Neutral

A close examination on the use of ICT by faculty suggests that majority of students were not satisfied with the extent of ICT used by the faculty for teaching. Irrespective of the background of students, 41.7 % of students were of the view that ICT used by faculty is inadequate. It can be inferred from the findings that faculty needs to use more ICT led teaching tools for the benefit of students. Opinion of students on the availability of ICT infrastructure was also sought. The distribution of their opinion is presented in Table 7. It can be seen from the table that highest percentage (56.3 %) of students were not satisfied with ICT infrastructure in the institutions. In fact majority (93.8 %) of urban background students were dissatisfied with ICT infrastructure. The lower level of dissatisfaction among rural background students could be attributed to the lesser use of ICT by those students. The association between level of dissatisfaction regarding ICT infrastructure and background of students is highly significant (1 % level).

Table 8 Initiatives to improve use of ICTs in higher education

Initiatives	No.	Percentage
Faculty needs to be motivated to use more ICT for teaching	44	61.1
Government should provide ICT infrastructure in the institution	39	54.2
Cyber-café should be set up in the premises on PPP model	46	63.9
There should be compulsory paper in ICT use for every student on the lines of environmental paper[a]	52	72.2

[a]It may be mentioned that CSJMU has environmental paper that must be cleared by every student before awarding any degree by the University though marks obtained in the paper are not included in the pass percentage

The opinion of students was also sought on the possible initiatives that can increase the usage of ICT by students of higher education. The distribution of their opinion is presented in Table 8.

It can be seen from the table that fairly large percentage (72.2 %) of students was of the view that there needs to be a compulsory paper on the use of ICT on the lines of environmental paper which is expected to accelerate the learning momentum among students. The second initiative in the opinion of sample students was to setup cyber-café in the premises of institution based on Public-Private-Partnership (PPP) model. Roughly 64 % of students suggested setting up cyber-café on PPP model. More than sixty (61.1 %) of students felt that faculty needs to be motivated for use of ICT led teaching methods. In other words students felt the need of regular faculty upgradation programmes that are prevalent in several universities in India and abroad.

4 Summary and Conclusion

The study aims at examining impact of Information and Communication Technologies (ICT) on higher education and to identifying major impediments that have resulted in slow penetration of ICT in higher education institutions in rural India. An attempt was also made to identify and analyse predicaments that hinder the diffusion of ICT in higher education institutions in rural areas. The paper is based on primary data collected from four higher education institutions located in Sitapur district, one of the six e-districts identified by the Government of Uttar Pradesh. The data from 72 students were collected through a semi-structure questionnaire. The survey was conducted during October 2012 and January 2013. Majority of the sample institutions were affiliated to Chhatrapati Sahu Ji Maharaj University, Kanpur.

While praising the initiatives taken by the University for digital Student-University Interface, findings of the study suggest that major impediment in the use of ICT in higher education institutions in rural areas has been the lack of ICT infrastructure. Most of the students were using digital technologies for downloading enrolment form, examination form, results and downloading reading material. The Students Login facility which has been provided by the University recently is a step forward in digital student-university communication. The findings suggest that there exists a digital divide among students of rural and urban background. Results also show that lack of awareness of faculty and lack of appreciation by students have also been reasons for lesser use of ICT led teaching methods by the faculty.

The study brings out empirical evidence to suggest that there is need for regular faculty upgradation programme and students also need to be educated about ICT led teaching tools. Based on the findings of the study it is recommended that cyber-café may be set up in the premises of higher education institutions so that students can make better use of their time and learn more about ICT. These cyber-cafés may be set up on Public-Private-Partnership (PPP) model. It is also recommended that there should be a compulsory paper at the graduation level on the use of ICT. This will ensure that any student passing out from the University is well aware of digital technologies.

Acknowledgments The authors would like to place on record the appreciation of Er. Ashish Verma, Manager of Shri D.P. Verma Memorial Degree College, who supervised the survey and to the staff of all four institutions who conducted the survey. They would not have been able to complete the paper without statistical assistance of Mr. Piyush Kumar. Most importantly special thanks are due to the students who willingly participated in the survey and expressed their free opinion.

References

Brown, W. B., & Liedholm, C. E. (2002). Teaching microeconomic principles—Can web courses replace the classroom in principles of microeconomics? *American Economic Review*, 444–448. May 2002 (Papers and Proceedings).

James, P., & Hopkinson, L. (2009). *Sustainable ICT in further and higher education.* Joint Information Services Committee Report, University of Bradford. http://www.ictliteracy.info/rf.pdf/rptgreenictv1.pdf. Accessed September 20, 2010.

Lal, K., & Paul, Shampa. (2004). Digital divide and international orientation: Evidence from Asia-Pacific countries. *ICFAI Journal of Applied Economics, 3*(2), 31–41.

Lal, K. (2007). *Information and communication technologies in the context of globalisation: Evidence from developing countries*. New York: Palgrave Macmillan.

Loing, B. (2005). *ICT and higher education.* IXth UNESCO/NGO collective consultation on higher education, http://www.unesco.org/ngo/comite/cpmother/enseign-sup/tic-gb.pdf. Accessed September 22, 2010.

Using Technologies to Reduce Unequal Access in Teacher Training

Kalpana Kannan and K. Narayanan

1 Introduction

Most developing countries around the world have shortage of qualified and trained teachers. We need better teachers and in large numbers to cater to the growing needs of the knowledge based society. Some of the challenges faced today in the education system, such as expanding the reach of education, imparting quality education at affordable costs and training a large number of teachers to address the shortage of qualified and experienced teachers is difficult to achieve using the traditional face-to-face method. We need a new approach to address these challenges. Implementation and integration of information and communication technology (ICT) into education system can address some of these challenges.

ICT enabled education is an emerging field. It offers both opportunities and challenges for researchers and educationalists. It fulfills various needs of the society which is otherwise difficult to achieve through the conventional education. Literature shows that many universities across the world have adopted ICT enabled education for higher education, skill development and professional development.

This chapter is an extension of the authors' work "Synchronous Teacher Training in India: A Study of Perceptions and Satisfaction of the Participants" that was published in 2015 in the *International Journal of Information and Education Technology* (IJIET), Vol. 5, No. 3.

K. Kannan (✉)
Department of Computer Science and Engineering,
Indian Institute of Technology Bombay, Mumbai, India
e-mail: kalpanak@cse.iitb.ac.in

K. Narayanan
Department of Humanities and Social Sciences,
Indian Institute of Technology Bombay, Mumbai, India
e-mail: knn.iitb@gmail.com

N.S. Siddharthan and K. Narayanan (eds.), *Technology*,
India Studies in Business and Economics,
DOI 10.1007/978-981-10-1684-4_11

Research shows that ICT enabled education has a positive impact on teachers and learners (Williams 2005; Looi et al. 2004).

In India, there has been a rapid expansion of institutions of higher education since independence in 1947. The number of universities and colleges has increased from 20 and about 500 in 1947 to approximately 700 and 20,000 respectively today (UGC 2011). Every year about 1 million undergraduate students graduate from various engineering disciplines (AICTE 2012). There are about 5000 engineering colleges and about 200,000 teachers teach in these colleges (AICTE 2012). Unfortunately, the quality of skill sets imparted in undergraduate education is low and many students do not find appropriate jobs. The quality of engineering education has suffered due to a variety of reasons: (a) The sudden expansion of colleges has led to a shortage of qualified and experienced teachers, (b) Low salary in the teaching profession has further added to the shortage, especially in the unaided colleges, and (c) Lack of adequate infrastructure in colleges (Kannan and Narayanan 2012).

The biggest resource crunch in higher technical education today is the non-availability of experienced and qualified teachers. To raise the level of engineering education in the country is a national level problem, for which there is no one solution. It has been observed that many engineering colleges in the country have a very few faculty with doctoral or even Master's degrees. In many colleges, fresh undergraduates join as a faculty member with no teaching experience. These young teachers usually do not feel encouraged to build a long-term academic career due to low salaries and the poor academic environment in the colleges. Hence, the quality of education in most of the engineering colleges is badly affected. Given the large number of engineering colleges and teachers in the country, the existing mechanism of training teachers through the Quality Improvement Program (QIP) where about 40–50 teachers are trained at a time is not scalable. Some of the limitations of the conventional training are: (a) Time and resources required are enormous, (b) Infrastructure constrains, (c) Limited number of participants can be accommodated, (d) College management may invariability choose regular senior faculty and (e) Training is mostly restricted to urban population. Some of these factors lead to unequal access for teacher training. One of the ways to provide equal access and to scale-up the professional development activity is through ICT enabled teacher training.

Use of ICT for synchronous teacher training enables the trainer to reach out to a large number of participants at a given point of time. Covering a large number of teachers is an important mechanism through which the quality of engineering courses can be improved. However, merely making the course available may not achieve these objectives. The mechanism used by an institution engaged in synchronous teacher training can help achieve this. Given the large diversity in the kind of training given to the teachers, it is important to ensure high quality with low cost and large outreach.

Historically, subjects of study have witnessed more and more male students joining engineering in general and mechanical engineering in particular. Introduction of Computer Science and Engineering branches provided an opportunity for a large number of female students to enroll. However, technological change and growth of plastics, fibers and rubber eased the bottleneck of traditional subject such as mechanical engineering for female students. The gender gap in student enrollment gets reflected in the availability of faculty as well in different courses. Very often we find that female teachers find it difficult to participate in training if conducted very far away from their place of work or home. The synchronous mode teacher training could be a source through which such issues can be resolved.

Diversity in the background and location could restrict the participants from gaining equally. This may especially happen when the training programme includes a large number of participants spread wide-over. The teacher training programme could therefore not be an ideal example to reap the economies of scale in terms of benefits or satisfaction. Does this hold true in the case of synchronous mode teacher training?

There are several initiatives undertaken by the government, both at the centre and the state level to facilitate skill upgradation of college teachers in technical institutions in India. In this study, we have collected data from the large scale synchronous distance mode teacher training for engineering college teachers, undertaken by the Indian Institute of Technology Bombay (IITB), under the National Mission on Education through ICT (NMEICT 2013), sponsored by the Ministry of Human Resource Development, Government of India. Specifically, this study assesses the impact of synchronous mode teacher training undertaken by Indian Institute of Technology Bombay (IITB) during the period 2009–2013. In order to train a large number of engineering college teachers in the country, the 'Train 10,000 Teachers Training (T10KT 2014)' programme at the Indian Institute of Technology Bombay (IITB) uses an Internet based software called AVIEW (2014) to conduct training workshops. These workshops are conducted in synchronous distance mode. The lectures are transmitted live from a single location (IIT Bombay) and 50–80 participants located at each of the 250 different geographical locations called remote centres (RCs) attend these workshops in synchronous remote classrooms. This hub and spoke training model is unique as all three forms of interactions happen—Synchronous interactions between the hub and RCs through AVIEW and chat; asynchronous interactions between instructors, participants and coordinators through discussion forum (Moodle) and email; and face-to-face interactions between participants and the coordinators in all RCs. Hence, this model can also be referred to as a blended synchronous mode training model (Kannan and Narayanan 2015a).

The objectives of our study were to:

- *investigate whether the synchronous mode teacher training programme helps achieve wide outreach and inclusivity.*
- *identify the differences, if any, that exist among the participating teachers of Computer science and Mechanical engineering workshops.*

- *examine the changes in the demographic characteristics, outreach, participants' perceptions and satisfaction level, when the training programme is scaled up.*

In order to study the outreach, inclusion, perceptions and satisfaction of participants of synchronous mode teacher training workshops across various disciplines, data were collected over a period of time from December 2009 to May 2013. With the help of online questionnaire, information on demography of participants, location of colleges, participants' feedback on a few critical factors viz., quality of instruction, usefulness of workshop content, quality of audio-video transmission and level of satisfaction were collected and analysed.

2 Literature Review

Most developing countries do not have access to ICT infrastructure. There is a digital divide. The digital divide is the gap between those who have access and control of technology and those who do not have the access. This means that integrating ICTs at different levels and various types of education is going to be a challenging task. Failure to meet the challenge would mean further widening of the knowledge gap and deepening of economic and social inequality (Tinio 2010). Improving the quality of education and training is a critical issue particularly in the developing countries. ICTs can enhance the quality of education in several ways—by increasing learner motivation and engagement, by facilitating the acquisition of basic skills and by enhancing teacher training.

Some of the developing countries have adopted ICT tools to improve the gross enrollment ratio in higher education (Sridhar 2005), teacher training (UNESCO 2001) and distance education through television and radio (Dixit 2009). The application of ICT has also revolutionized and enriched education and skill development in the rural areas (Ramos et al. 2007). People with disabilities (Taylor 2005; Wild 2004), and women who could not earlier participate in learning process due to socio-cultural barriers now can (Sridhar 2005). ICTs are powerful tools for extending educational opportunities outside the boundaries of traditional classrooms. With the help of ICT, one can reach out to a larger population—rural population, women, minority groups, people with disabilities, elderly, and those who are unable to participate in the learning activity because of cost or lack of time. Thus, with the help of ICTs one can reduce the inequalities in the society.

Educational researchers propose several benefits of using ICTs for educational purposes. Synchronous learning approaches enables equity of access for learners who are geographically isolated and it also has better completion rate than asynchronous (Norburg 2012). Synchronous mode also promotes discussion and community learning (Roseth et al. 2013; Lidstone and Shield 2010).

2.1 Use of ICT in the Indian Context

In India, the need to use technology in education started mainly to reach out to a larger population and also to improve the quality of education at all levels. Arora and Pandey (1998) had analysed the effectiveness and usefulness of distance education/training for a large number of school teachers, conducted by the National Council for Educational Research and Training (NCERT) in collaboration with Indira Gandhi National Open University (IGNOU 2014). According to the authors, the use of interactive video technology was found to be suited for training of teachers especially in remote parts of the country; they also suggested that the training programme through distance mode should become a frequent feature to make it cost-effective. Bhattacharya (2008) reviewed some of the technology enhanced initiatives such as, National Programme on Technology Enhanced learning (NPTEL 2014), use of educational satellite (EDUSAT 2013), virtual classrooms and virtual laboratory undertaken by the Government of India to improve the quality of engineering education in India.

2.2 Benefits of Using ICT for Teacher Training

1. Cost to employer: In the conventional mode, there are high costs associated with travel and accommodation. These costs add up to the training costs of the employer/state. The cost of training can be significantly reduced by using distance mode of training (Berge 2007).
2. Lack of time for teachers: Most teachers find it difficult to stay away from work for long period of time for training purpose. Therefore, if the training is provided closer to their work place, it saves time (Berge 2007).
3. Reaching a large number in short time: Imparting uniform training to a large number of people spread across geographical areas is costly and difficult. Distance training with the help of ICT can make it both economical and convenient to the trainer and the trainee (Berge 2007).
4. Doing more with less: Even though training is considered essential, most organizations do not have enough budgets for training. Therefore, using distance mode one can reach out to a large number of teachers with limited budget (Berge 2007).

2.3 Studies on Teacher Training Using ICT

Many studies have been conducted in various countries on teacher training using ICT. Table 1 gives the summary of some of these studies. The results indicate that

Table 1 Studies on teacher training using ICT

Author and year	Country	Purpose	Findings
Wang et al. (2010)	Australia	The study proposed a training model called Practice, reflection and collaboration and discussed how it was implemented for Chinese language teacher in cyber environment	Collaborative learning proved to be an important component in training and in supporting learning communities
Adem (2009)	Ethiopia	The study describes the distance education teacher training programmes it has been running since the inception of IICBA in Sub-Saharan Africa through the use of modern pedagogy and ICT	The study found that some of the drawbacks of the programme were: lack of accreditation and support from distance educational institution, lack of incentive, lack of time to study along with regular job which lead to high dropout rates
Young and Lewis (2008)	USA	To study the perception of pre-service teachers towards a training program delivered at a distance	Teachers had a positive response to distance education in terms of overall satisfaction. Female teachers were more concerned about the student-teacher and peer-to-peer interaction than male teachers
Zhang and Hung (2007)	China	To determine if there was an improvement in school teachers in pedagogy and linguistic competencies by using ICT	Teaching and language skills of teachers improved after the training. There was a low dropout rate (3 %)
Pouezevara and Khan (2007)	Bangladesh	To find perceptions and attitude on content, training effectiveness in using mobile technology for in-service secondary school teachers in rural Bangladesh	Distance mode teacher training can be as effective as face-to-face. It also helps to build learning communities

there are many benefits of using ICT for teacher training, especially if one needs to address the unequal access to teacher training and teaching material.

3 Objectives and Methodology

The objectives of our study were:

(a) *To investigate whether the synchronous mode teacher training programme helps achieve wide outreach and inclusivity.*

The outreach of the training workshops was measured in terms of the number of states covered, number of cities/towns, number of colleges and number of participants who attended the workshops through various remote centres from 2009 to 2013. Inclusion included parity in terms of rural/urban, gender, age group and educational qualification.

(b) *To identify the differences, if any, that exist among the participating teachers of Computer science and Mechanical engineering workshops.*
We compared the gender, age group, educational qualification and location of participating teachers in three workshops in the area of Computer science and engineering and three workshops in the area of Mechanical engineering. We also compared participants' perceptions and levels of satisfaction in order to find out the differences between these two sets of workshops.

(c) *To examine the changes in the demographic characteristics, outreach, participants' perceptions and satisfaction level, when the training programme is scaled up.*
In order to examine the changes in the demographic characteristics, outreach, participants' perceptions and satisfaction level when the training programme is scaled up, two workshops conducted by the same instructor with two different class sizes were compared. One workshop on Database Management System (DBMS) was conducted in the 1000 teachers training programme and the same workshop was repeated after a gap of two and a half years under the 10,000 teachers training workshop by the same instructor. Similarly, two workshops in the subject 'Thermodynamics' was conducted by the same instructor. First workshop was conducted under the 1000 teacher training programme and the second workshop was conducted after a gap of one and a half years for a larger number of teachers under the 10,000 teachers training programme.

4 Data Collection

As the number of participants was large and geographically scattered, online survey method was adopted to collect data. An instrument was developed to collect data on demography and to measure other variables in the study. After every workshop, online questionnaire was posted on the workshop learning management system 'Moodle' and participants were given ten days time to fill the questionnaire. Information on gender, age group, educational qualification, work experience and location of colleges of participants in rural/urban areas were collected through the questionnaire. Other items to measure participants' perceptions on the quality of instruction, usefulness of course content, audio and video quality were measured using a five-point Likert scale from 'Strongly disagree = 1' to 'Strongly agree = 5'. Literature shows that Likert scale is a popular method used by researchers (Young and Lewis 2008; Davis 1989) to measure people's attitudes, perceptions,

Table 2 Sample size and response rate

W. no	Workshop title	Number of participants	Sample size (N)	Response rate (%)
1	Computer programming (Pilot)	637	463	73
2	Computer programming	641	448	70
3	Database management system (DBMS)	1022	710	70
4	Thermodynamics	779	686	88
5	Basic electronics	1267	1029	81
6	Software development techniques	773	502	65
7	Heat transfer	735	707	96
8	Solar photovoltaics	930	845	90
9	Computational fluid dynamics (CFD)	1253	1001	79
10*	Introduction to research methods (RM)	7506	4173	55
11*	Engineering thermodynamics	4759	2981	62
12*	DBMS	7423	3703	50

Note *Workshops 10, 11 and 12 were conducted under the Train 10,000 teachers training programme (T10KT)

preferences, opinions etc. Therefore, Likert scale was chosen to measure the items used in the study. The overall satisfaction was measured using the scale 'Poor = 1', 'Satisfactory = 2', 'Good = 3' and 'Excellent = 4'.

These workshops had a very large sample size. On an average about 70 % of the participants responded to our online questionnaire. The sample size for the initial workshops were about 450 and later as the workshop was scaled up to ten times the sample size also increased to about 3700. Table 2 gives the details on sample size and response rate for all the workshops.

5 Analysis and Discussion

(a) *To investigate whether the synchronous mode teacher training programme helps achieve wide outreach and inclusivity.*

We measured the outreach of the training workshops in terms of the number of states covered, number of cities/towns, number of colleges and number of participants who attended the workshops through various remote centres from 2009 to 2013. Table 3 gives the statistics on workshops conducted from December 2009 to May 2013. After analyzing the data we observe that these workshops have a very good outreach. It can be noted from the data given in Table 3 that the number of

Table 3 Workshop statistics

W. no.	Period	Workshop	Participants	RCs	States	Cities/towns	Colleges
1	Dec 2009	Computer programming (Pilot)	637	22	6	204	70
2	July 2010	Computer programming	641	30	13	218	130
3	Dec 2010	Database management system (DBMS)	1022	32	13	158	161
4	June 2011	Thermodynamics	779	31	7	248	424
5	June 2011	Basic electronics	1267	38	12	401	429
6	Nov 2011	Software development techniques	773	21	10	113	300
7	Dec 2011	Heat transfer	735	28	14	377	327
8	Dec 2011	Solar photovoltaics	930	35	14	245	350
9	June 2012	Computational fluid dynamics (CFD)	1253	47	9	275	561
10*	June 2012	Introduction to research methods (RM)	7506	168	19	560	2000
11*	Dec 2012	Engineering thermodynamics	4759	167	18	701	1209
12*	May 2013	DBMS	7423	245	21	801	1121

RCs Remote centres
*Workshops 10, 11 and 12 were conducted under the Train 10,000 teachers training (T10KT) programme

remote centres were scaled up from 22 to 245 centres, the number of states covered during the workshop went up from 6 to 21, the number of participants went up from 637 to about 7423, the number of cities/towns went up from 204 to 801, the number of colleges covered went up from 70 to 1209. Thus, these workshops could be easily scaled up to ten times in three years.

Figure 1 shows the growth and outreach in terms of the number of participants, the number of colleges and the number of centres.

Inclusion included parity in terms of gender, age group, rural/urban and educational qualification. Tables 4 and 5 give the demographic characteristics of the participants for all the workshops. The gender distribution is given in Fig. 2. We observe that except for mechanical engineering courses, most of the workshops had about 40 % female participants. In mechanical engineering courses the percentage of female participants was about 12 %. This could be because the number of female studying mechanical engineering in the country is few. One of the reasons cited for a large number of female participation in these workshops was proximity of the

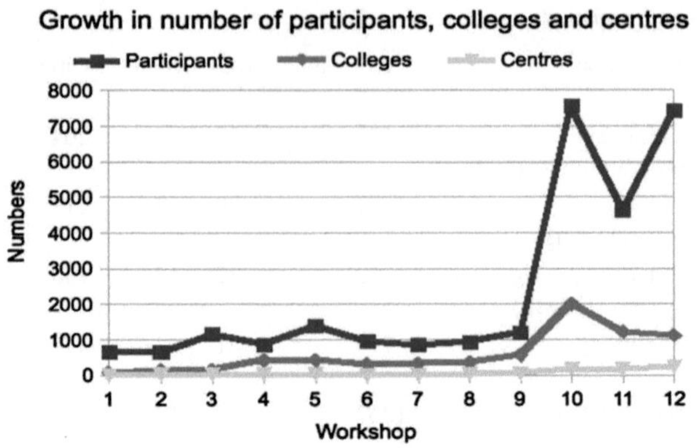

Fig. 1 Growth in the number of participants, colleges and centres

Table 4 Gender, age group and location of participants

Workshop	Sample N	Gender (in %)		Age group (in %)			Location (in %)	
		Male	Female	23–30	31–40	Above 40	Urban	Rural
Computer programming (Pilot)	463	48	52	65	29	6	65	35
Computer programming	448	51	49	68	26	6	58	42
DBMS	710	53	47	65	29	6	58	42
Thermodynamics	686	88	12	46	38	18	55	45
Basic electronics	1029	55	45	57	32	11	56	44
Software development techniques	502	63	37	62	33	5	57	43
Heat transfer	707	88	12	43	41	16	55	45
Solar photovoltaics	845	69	31	42	40	18	53	47
CFD	1001	87	13	43	43	14	62	38
Introduction to RM	4173	60	40	49	39	12	57	43
Engineering thermodynamics	2981	87	13	57	31	13	51	49
DBMS	3703	53	47	63	32	5	57	43

remote centres. Some of the female participants told us that because of the proximity of the remote centre it was easier for them to attend the workshops. They did not have to leave their families for a long period of time to attend the workshops (Kannan and Narayanan 2015b). Most of them could travel from home to the remote centres everyday, like the way they would be to their colleges. Not disturbing their family life was one of the major factors that enabled female teachers and others with elderly members in the family to attend the workshop.

Table 5 Educational qualification and work experience of participants

Workshop	Sample (N)	Edu. Quali. (in %)			Work experience (in %)			
		UG	PG	PhD	<2 years	2–5 years	5–10 years	>10 years
Computer programming (Pilot)	463	NA	NA	NA	NA	NA	NA	NA
Computer programming	448	30	67	3	NA	NA	NA	NA
DBMS	710	32	65	3	NA	NA	NA	NA
Thermodynamics	686	29	61	10	21	28	22	29
Basic electronics	1029	35	59	6	21	31	27	21
Software development techniques	502	21	75	4	20	32	31	17
Heat transfer	707	20	69	11	18	28	24	30
Solar photovoltaics	845	15	71	14	20	20	25	35
CFD	1001	16	75	9	19	27	4	30
Introduction to RM	4173	12	81	7	18	28	29	25
Engineering thermodynamics	2981	27	66	7	25	33	20	22
DBMS	3703	18	78	4	23	39	27	11

NA Data not available

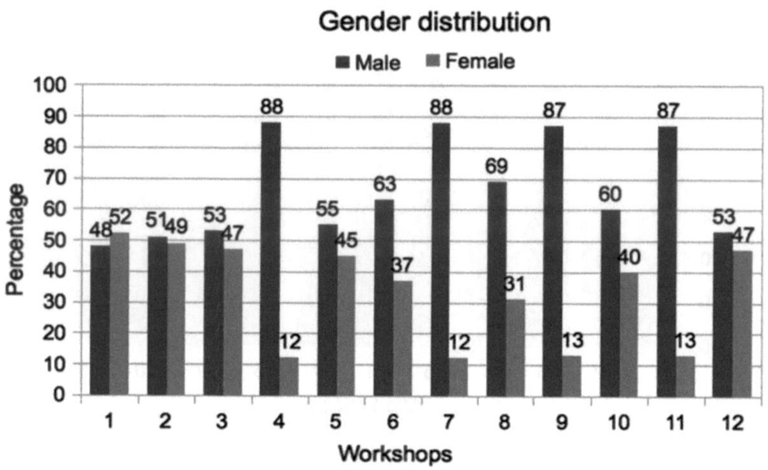

Fig. 2 Gender distribution

The age distribution is given in Fig. 3. From the data we observe that about 60 % of the participants were below 30 years and about 40 % of the teachers were above 30 years of age. The benefits of the workshops would be more if the majority of teachers are young as they can experiment with new ideas and implement it in their classroom teaching. It is beneficial to catch them young in order to have

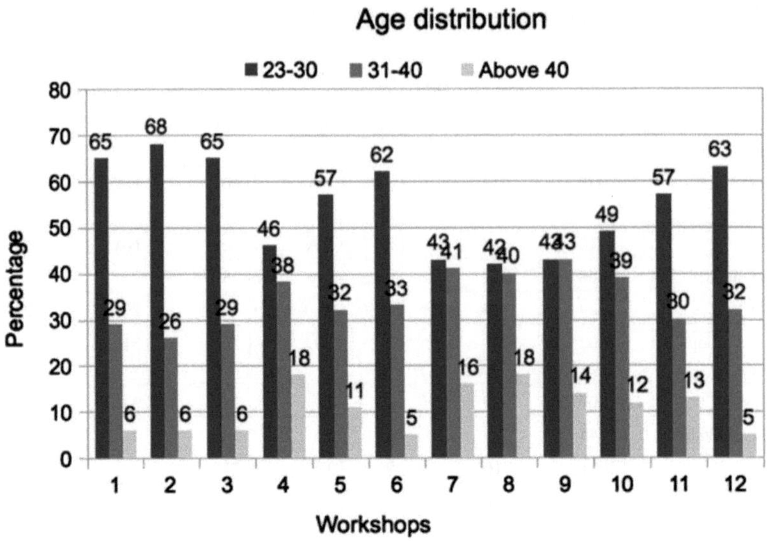

Fig. 3 Age distribution

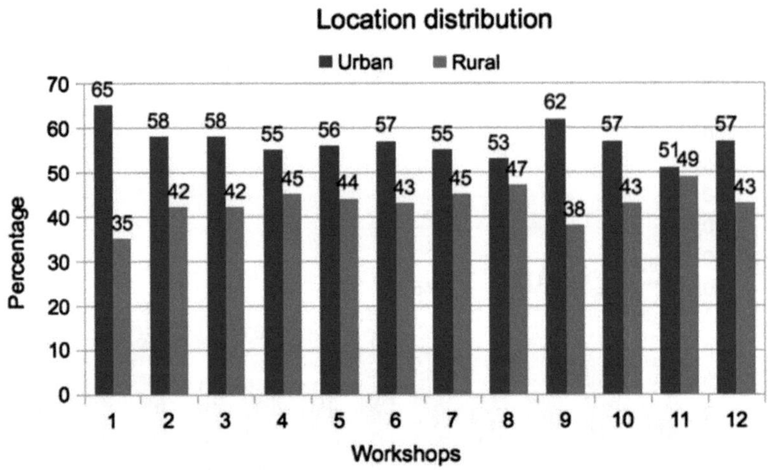

Fig. 4 Location distribution

maximum impact. In most teacher training programme conducted in the conventional face-to-face mode, preferences are normally given to senior staff members, keeping in mind their long services. However, given that synchronous teacher training can accommodate more participants, younger faculty everywhere gets a good chance to attend. The requirement of University Grant Commission (UGC) also gets fulfilled.

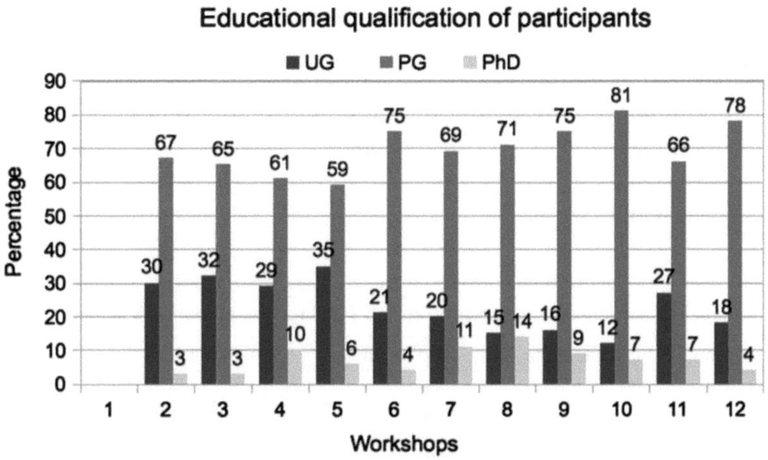

Fig. 5 Educational qualification of participants

In all the workshops, about 40 % of teachers came from colleges located in the rural areas. The location distribution is given in Fig. 4. One of the advantages of technology-enabled workshops is reaching out to remote places in the country. We also observed that as the number of remote centres went up from 22 to 245, the gap between the rural and the urban decreased. More number of teachers from the rural engineering colleges could participate in the workshop after scaling up. More publicity, easier accessibility to remote centres could also be the reason for this.

Figure 5 gives the educational qualification of the participants. On an average, about 70 % of the respondents were postgraduates, about 23 % were undergraduates and about 7 % were doctorates.

Table 5 gives the data on the educational qualification and work experience of the participants. On an average, about 75 % of the respondents had less than 10 years of work experience and about 25 % had more than 10 years of work experience. This data also reveals that majority of the participating teachers were young and had less than 10 years of work experience. Thus, we can conclude that in the synchronous mode teacher training programme a lot of young teachers are benefitted compared to the conventional face-to-face mode, where the majority could be senior teachers, as they get the preference due to the limited numbers of seats.

In order to compare the outreach and inclusion of synchronous mode with the conventional face-to-face Quality Improvement programme (QIP), we collected data on the QIP short term courses conducted by IIT Bombay during the same time period from 2009 to 2013. Table 6 gives the data on the number of QIP courses conducted, the number of participants and the average number of participants per course.

From Table 6, we observe that in the conventional face-to-face QIP courses the average number of participants per course was about 27 participants, which is far

Table 6 QIP short term courses conducted by IIT Bombay (2009–2013)

S. no.	Period	Total number of courses	Total number of participants	Avg. no. of participants per course
1	2009–2010	8	175	22
2	2010–2011	9	208	23
3	2011–2012	10	257	26
4	2012–2013	9	245	27

Table 7 Level of participant's satisfaction for all synchronous mode workshops

Workshop no.	Excellent (in %)	Good (in %)	Satisfactory (in %)	Poor (in %)
1	52	35	12	1
2	43	52	5	0
3	51	44	4.7	0.3
4	47.7	44	8	0.3
5	31	54.5	14	0.5
6	36	56	7.4	0.6
7	45	48	7	0
8	49	44	6.7	0.3
9	36	49.4	14	0.6
10	35	53.6	11	0.4
11	33	48	18.2	0.8
12	40	47	12	1

less than the synchronous mode teacher training workshops. It was also observed that over the years the average number of participants had remained constant without scaling up. The conventional method surely cannot fulfill the existing need for quality improvement of college teachers.

Thus, after analyzing the data we conclude that the issues of outreach and inclusion are well addressed by the synchronous mode teacher training workshops which may be very difficult to achieve in a conventional face-to-face training. Reaching out to a very large number of teachers in a short time and inclusion in terms of gender, age group, location and educational qualification is easily possible through these synchronous mode workshops. This is strongly supported by the analysis of data collected from all the workshops. The level of satisfaction of participants for all the synchronous mode workshops is given in Table 7. The data shows that about 90 % of participants rated their experience as excellent or good.

(b) *To identify the differences, if any, that exist among the participating teachers of Computer science and Mechanical engineering workshops.*

In order to identify the differences between the two sets of courses, we compared the gender, age group, educational qualification and location of participating teachers in three workshops in the area of Computer science and engineering (CS) and three workshops in the area of Mechanical engineering (ME).

The workshops in the area of CS were Computer programming, Database Management System and Software Development Techniques. The workshops in the area of ME were Thermodynamics, Heat Transfer and Computational Fluid Dynamics.

It is the set of courses for two streams of engineering, Mechanical and Computer science, to identify the differences, if any, among the beneficiaries of the training programme. The level of education, age, gender and work experience may significantly vary among the participating teachers in these two engineering disciplines. These differences may get reflected in the impact analysis, especially in terms of coverage/outreach and levels of satisfaction. This is the reason why, a comparison was attempted. Statistics on the number of participants, the number of remote centres, cities/towns and the number of colleges represented in CS and ME workshops are given in Table 8. The demographic data is tabulated in Table 9 and

Table 8 Computer science and mechanical engineering workshop statistics

No.	Month-year	Workshop	Participants	Remote centres	Cities/ towns	Colleges
Computer science						
1	July 2010	Computer programming	641	30	218	130
2*	Dec 2010	Database management system (DBMS)	1022	32	158	161
3	Nov 2011	Software deveopment techniques	773	21	113	300
Mechanical engineering						
4*	June 2011	Thermodynamics	779	31	248	424
5	Dec 2011	Heat transfer	735	28	377	327
6	June 2012	Computational fluid dynamics (CFD)	1253	47	275	561

Note Workshop marked * were used for comparison with large scale workshops

Table 9 Demography of participants

CS and ME workshop	N	Gender (in %)		Age group (in %)			Location (in %)		Ed. Quali. (in %)		
		M	F	23–30	31–40	>40	Urban	Rural	UG	PG	PhD
C. prog.	448	51	49	68	26	6	58	42	30	67	3
DBMS	710	53	47	65	29	6	58	42	32	65	3
S/W dev.	502	63	37	62	33	5	57	43	21	75	4
Thermo	686	88	12	46	38	18	55	45	29	62	10
HT	707	88	12	43	41	16	55	45	20	69	11
CFD	1001	87	13	43	43	14	62	38	16	75	9

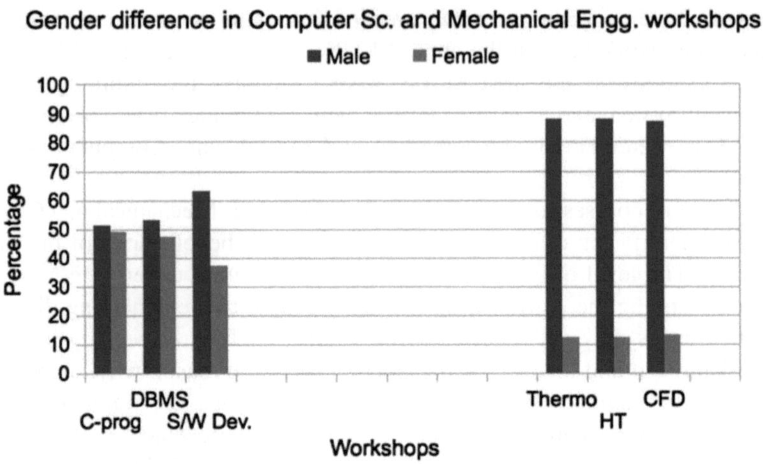

Fig. 6 Gender differences among the participants of CS and ME workshops

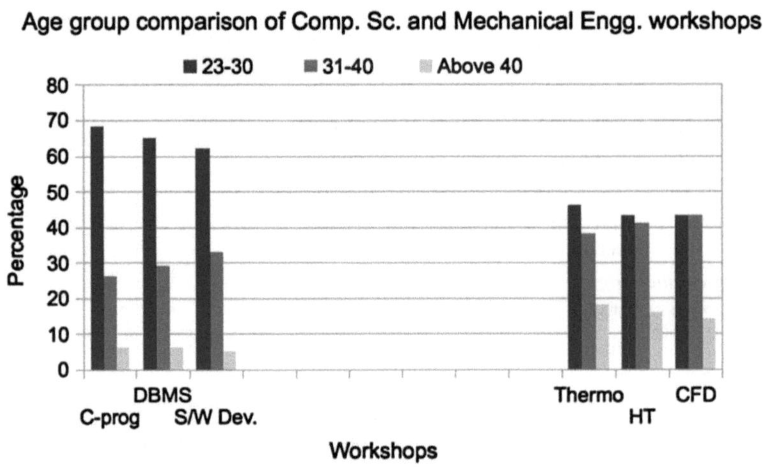

Fig. 7 Age group comparison of participants of CS and ME workshops

is graphically represented in Figs. 6, 7, 8 and 9. We also compared participants' perceptions and levels of satisfaction in these set of workshops.

The results indicate a clear difference in terms of gender, age and educational qualifications in CS and ME workshops. In CS workshops, on an average there were about 45 % female participants, whereas, in ME workshops the average number of female participants were 12 %. This difference could be mainly due to the fact that there are a very few female teachers in the area of Mechanical engineering. Hence, the number of female teachers in Mechanical engineering was less in comparison with the Computer science.

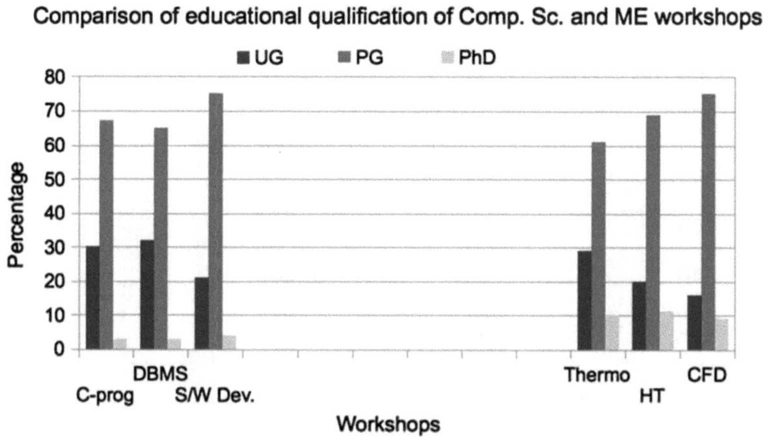

Fig. 8 Comparison of educational qualification of CS and ME workshops

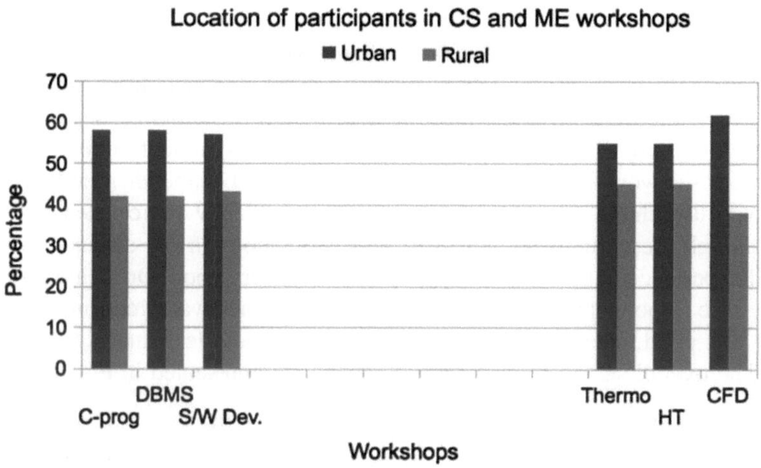

Fig. 9 Location of participants of CS and ME workshops

From Fig. 7 we observe that about 65 % of participants in CS workshops were in the age group 23–30 years, whereas, in ME workshops about 43 % were in the age group 23–30 years. The results indicate that a large number of young teachers participated in CS workshops in comparison with ME workshops.

From Fig. 8 we observe that in both CS and ME workshops maximum number of participating teachers about 65 % were postgraduates. However, we found that in ME workshops about 10 % were Doctorates, whereas, in CS about 3 % were Doctorates. The percentage of teachers with a Doctorate degree was higher in the conventional engineering discipline such as Mechanical engineering than in Computer science.

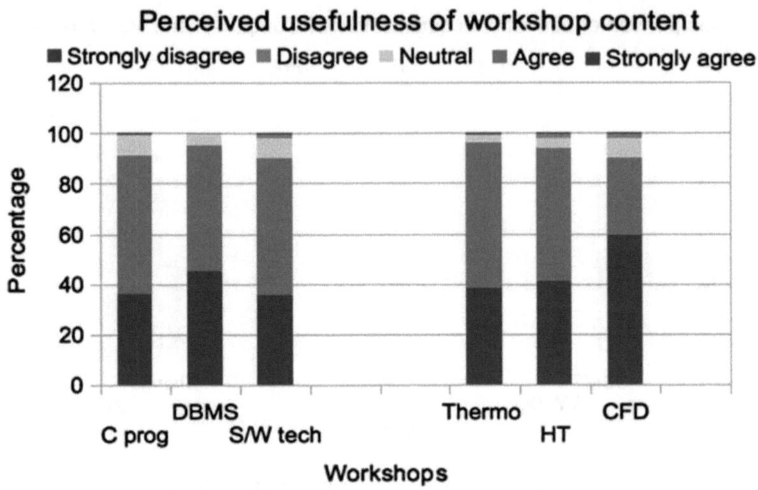

Fig. 10 Perceived usefulness of the workshop content

We found that on an average in both CS and ME the percentage of teachers who came from colleges located in urban area were about 60 % and from rural area were about 40 %. Figure 9 gives the location distribution of participants.

In both CS and ME workshops, about 90 % of the respondents agreed or strongly agreed that the workshop content was useful and would enable them to teach better. Perceived usefulness of workshop content is graphically represented in Fig. 10. Even though the participants attended these workshops in the distance mode, the overall experience of the participating teachers was very good. On an average, about 40 % of the respondents said that their overall experience was excellent, about 50 % said it was good, about 9 % said it was satisfactory and less than 1 % said it was poor. The overall satisfaction of the participants is represented in Fig. 11.

(c) *To examine the changes in the demographic characteristics, outreach, participants' perceptions and satisfaction level, when the training programme is scaled up.*

In order to examine the changes in the demographic characteristics, outreach, participants' perception and satisfaction level when the training programme is scaled up, we compared two workshops conducted by the same instructor with two different class sizes. We compared two Database Management System (DBMS) workshops, one with a small class size and another with a large class size. The first workshop was conducted in December 2010 with 1022 participants. Out of these, 710 participants responded to our survey. The second workshop was conducted by the same instructor after a gap of two and half years in May 2013. In this workshop, the number of participants was 7423, out of these, 3703 responded to our survey.

Similarly, we compared two Thermodynamics workshops, one with a small class size and another with a large class size. The first workshop was conducted in June

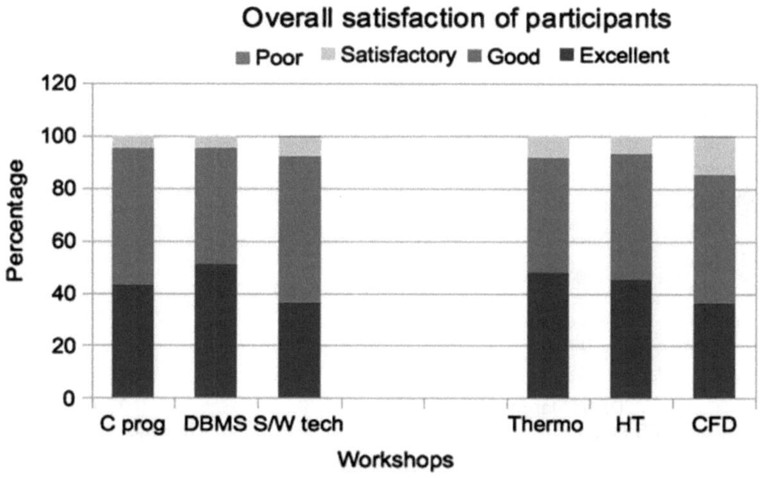

Fig. 11 Overall satisfaction of participants

Table 10 Workshop statistics of a small and a large workshop

No.	Period	Workshop	Participants	RCs	States	Cities/towns	Colleges
1	Dec 2010	DBMS (small)	1022	32	13	158	161
2	May 2013	DBMS (large)	7423	245	21	801	1121
3	June 2011	Thermodynamics (small)	779	31	7	248	424
4	Dec 2012	Thermodynamics (large)	4759	167	18	701	1209

Table 11 Demography of a small and a large workshop

Workshop	N	Gender (in %)		Age group (in %)			Location (in %)		Ed. Quali. (in %)		
		M	F	23–30	31–40	>40	Urban	Rural	UG	PG	PhD
DBMS (S)	710	53	47	65	29	6	58	42	32	65	3
DBMS (L)	3703	53	47	63	32	5	57	43	18	78	4
Thermo (S)	686	88	12	46	38	18	55	45	29	62	10
Thermo (L)	2981	87	13	57	31	13	55	49	27	66	7

DBMS (S) Database management system (Small), *DBMS (L)* Database management system (Large), *Thermo(S)* Thermodynamics (Small), *Thermo (L)* Thermodynamics (Large)

2011 with 779 participants. Out of these, 686 participants responded to our survey. The second workshop was conducted by the same instructor after a gap of one and half years in December 2012. The number of participants in this workshop was 4759, out of these, 2981 responded to our survey. The workshop statistics is given in Table 10. The demography of a small and a large workshop is given in Table 11.

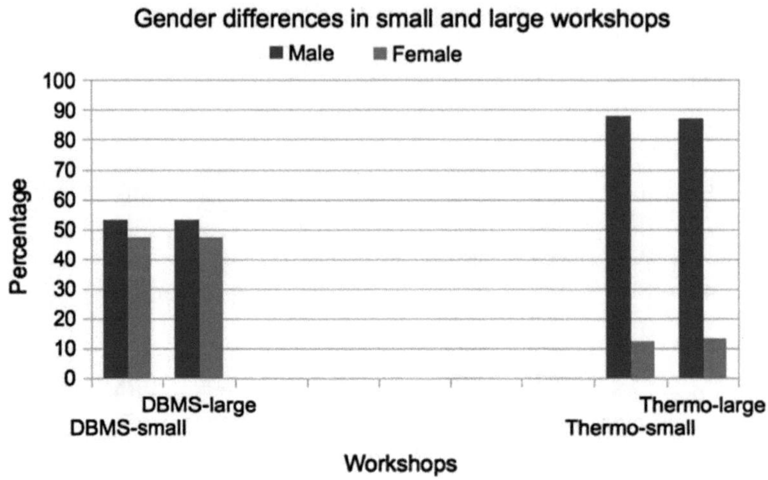

Fig. 12 Gender distribution in a small and a large workshop

A small and a large workshop in both DBMS and Thermodynamics were compared to see if there existed any difference in terms of gender, age group, location and educational qualifications. We found that even after scaling up the number of participants, the gender ratio remained the same. In both the DBMS workshops the percentage of male participants was 53 % and female participants was 47 %. Similarly, in both the Thermodynamics workshops the percentage of male participants was about 88 % and female was about 12 %. Figure 12 gives the gender distribution in a small and a large workshop.

The comparison of age group in the small and the large workshop shows that there was not much change in the composition in the DBMS workshop. Whereas, in the Thermodynamics workshop after scaling up the number of participants in the age group 23–30 years went up from 46 to 57 % and the percentage in the higher age group came down. The age distribution is graphically represented in Fig. 13.

We observed that when the workshops were scaled up, the number of participants from the rural engineering colleges also increased from 42 to 43 % in the case of DBMS workshop and from 45 to 49 % in the case of Thermodynamics workshops thus, reducing the gap between the urban and rural college teachers. This is a good indicator, as scaling up included more teachers from the rural engineering colleges, who have fewer opportunities than their urban counterpart. We also observed that in comparison with the DBMS workshop, in Thermodynamics workshop there were more number of participants from the rural engineering colleges. One of the reasons could be that Mechanical engineering is a traditional engineering subject that is offered in many colleges, even those located in the rural areas. The location distribution is given in Fig. 14.

We observed that after scaling up, in the DBMS workshop, the percentage of participants with the postgraduate degree increased from 65 to 78 %, and in

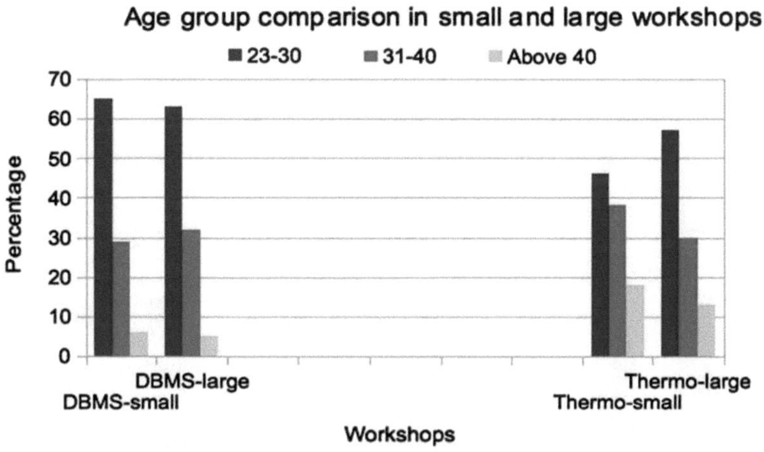

Fig. 13 Age group comparison in small and large workshops

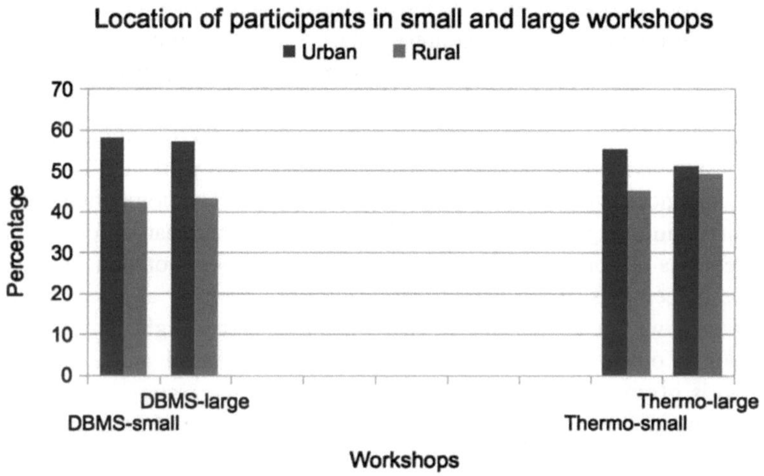

Fig. 14 Location distribution in small and large workshops

Thermodynamics workshop it increased from 62 to 66 %. The percentage of participants with an undergraduate degree decreased from 32 to 18 % in DBMS and 29 to 27 % in Thermodynamics workshop. The percentage of participants with a doctorate degree increased from 3 to 4 % in DBMS, whereas, it decreased from 10 to 7 % in Thermodynamics workshop. Figure 15 gives the graphical representation of educational qualification of participants.

We also observed that after scaling up, the outreach in terms of the number of participants, the number of remote centres, the number of cities/towns and the number of colleges increased by seven times in the case of DBMS workshop and it increased by six times in the case of Thermodynamics workshop. It seems it is

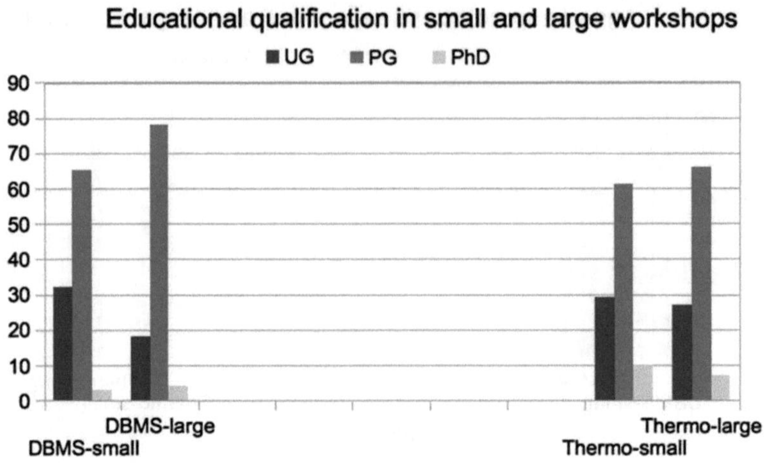

Fig. 15 Educational qualification in small and large workshops

relatively easier to scale up in CS related subjects. A possible reason could be that most engineering colleges in the country offer CS as a discipline as there is a higher demand due to the IT job market. The same may not be true in the case of Mechanical engineering.

We compared the perceptions of the participants on the quality of instructions, quality of audio/video interactions, and usefulness of the workshop content in two DBMS workshops with a small and a large class size. The comparison of participants' perceptions is given in Table 12. We found that the mean difference between the perceptions of participants in the two workshops were not statistically significant. The perceptions of the participants were not adversely affected due to scale up. However, we found that due to scale up there was a slight dip in the overall satisfaction. Comparison of overall satisfaction of a small and a large DBMS workshop is given in Table 13.

We compared the means of two workshops with the help of t-test and found that the difference was statistically significant. The results show that the participants of the smaller workshop (*Mean* = 3.45, *SD* = 0.6) were more satisfied in comparison with the larger workshop (*Mean* = 3.25, *SD* = 0.7). The mean difference between the two workshops was statistically significant ($t = 8$, $p \leq 0.001$).

Similarly, we compared the perceptions of the participants on the quality of instructions, quality of audio/video interactions, and usefulness of the workshop content in two Thermodynamics workshops with two class sizes. The comparison of participants' perceptions is given in Table 14. We found that some participants felt that there was slight reduction in the quality of audio and video quality after scaling up. We compared the mean of two workshops with the help of t-test and found that the difference was statistically significant ($t = 2$, $p \leq 0.05$). Comparison of overall satisfaction of the small and the large Thermodynamics workshop is given in Table 15. The results show that the participants of the workshop with the

Table 12 Comparison of participants' perceptions of two DBMS workshops

Statement	Sample	SA	A	N	D	SD	Mean	Std. Dev.	t-value
Instructor used innovative teaching methods to explain difficult concepts	710 (S) 3703 (L)	263 1481	404 1852	36 296	3 59	4 15	4.29 4.27	0.63 0.70	0.8 (n.s.)
The quality of audio and video transmission was good	710 (S) 3703 (L)	227 1851	376 1518	71 296	24 30	12 8	4.1 4.3	0.83 0.68	−0.87 (n.s.)
Workshop content was useful and would help in teaching	710 (S) 3703 (L)	320 1851	355 1666	31 148	3 31	1 7	4.39 4.43	0.6 0.39	−1.6 (n.s.)

Note Sample size DBMS (Small)—710, Sample size DBMS (Large)—3703
SA Strongly agree, *A* Agree, *N* Neutral, *D* Disagree, *SD* Strongly disagree

Table 13 Comparison of overall satisfaction of two DBMS workshops

	Sample	Excellent	Good	Satisfactory	Poor	Mean	Std. Dev.	t-value
Overall satisfaction	710 (S) 3703 (L)	362 1481	312 1740	33 444	3 38	3.45 3.25	0.6 0.7	8***

*** $p \leq 0.001$

Table 14 Comparison of participants' perceptions of two Thermodynamics workshops

Statement	Sample	SA	A	N	D	SD	Mean	Std. Dev.	t-value
Instructor used innovative teaching methods to explain difficult concepts	686 (S) 2981 (L)	329 1222	308 1490	28 209	14 50	7 10	4.30 4.29	0.74 0.69	0.3 (n.s.)
The quality of audio and video transmission was good	686 (S) 2981 (L)	219 894	391 1550	55 358	14 149	7 30	4.1 4.04	0.73 0.83	2**
Workshop content was useful and would help in teaching	686 (S) 2981 (L)	260 1103	3,981,669	20 179	4 23	4 7	4.32 4.28	0.61 0.62	1.6 (n.s.)

Note Sample size Thermodynamics (Small)—686, Sample size Thermodynamics (Large)—2981
SA Strongly agree, *A* Agree, *N* Neutral, *D* Disagree, *SD* Strongly disagree
**$p \leq 0.05$

Table 15 Comparison of overall satisfaction of two Thermodynamics workshops

	Sample	Excellent	Good	Satisfactory	Poor	Mean	Std. Dev.	t-value
Overall satisfaction	686 (S) 2981 (L)	326 984	303 1430	55 543	2 24	3.38 3.13	0.64 0.72	10.41***

***$p \leq 0.001$

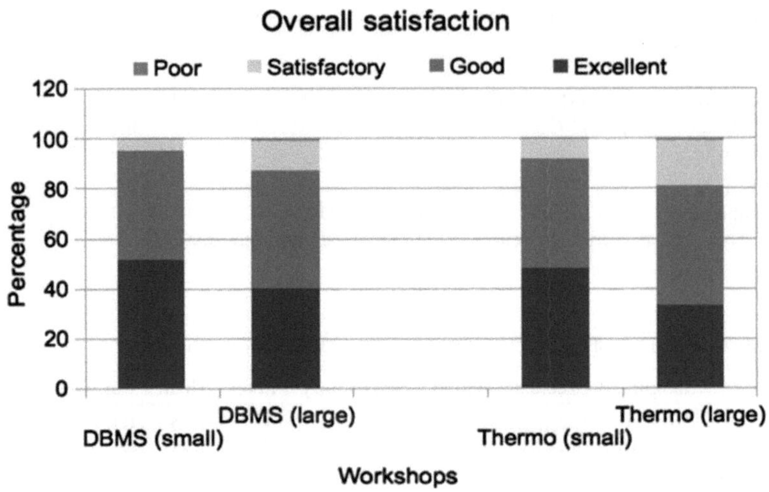

Fig. 16 Overall satisfaction of participants

small class size (*Mean* = 3.38, *SD* = 0.64) were more satisfied in comparison with the large class size (*Mean* = 3.13, *SD* = 0.72). The mean difference between the two workshops was statistically significant ($t = 10.41, p \leq 0.001$). The results were similar to the DBMS workshops.

During subsequent interactions with the participants, we found that one of the reasons for dissatisfaction among the participants was that many remote centres did not get an opportunity to interact live with the instructors and that caused some dissatisfaction. Also, the quality of audio/video was depended on the local set up and the Internet bandwidth. Therefore, there were some variations in the quality of delivery due to scaling up. Overall satisfaction of participants is graphically represented in Fig. 16.

6 Summary and Conclusions

The analysis of results indicate that the issues of outreach and inclusion were very well addressed by the synchronous mode teacher training workshops, which may be very difficult to achieve in the conventional face-to-face training. It was found that the outreach of these workshops was good in terms of the number of participants, number of cities/towns, number of states and colleges covered. Also, inclusivity in terms of gender, age group, location and educational qualification was well achieved. A large number of young teachers, female teachers and teachers in the rural colleges were benefitted from the synchronous mode teacher training workshops. It was possible to reach out to a very large number of teachers in a short time without drastically affecting the quality of instructions through these synchronous training workshops. This could have been very difficult to achieve through the traditional face-to-face training.

We compared the gender, age group, educational qualification and location of participating teachers in three workshops in the area of Computer science and engineering (CS) and three workshops in the area of Mechanical engineering (ME). The results indicate a clear difference in terms of gender, age and educational qualifications in CS and ME workshops. In CS workshops, on an average there were about 45 % female participants, whereas, in ME workshops the average number of female participants were 12 %. The results indicate that a large number of young teachers participated in CS workshops (65 %) in comparison with ME workshops (43 %). We found that in ME workshops about 10 % were Doctorates, whereas, in CS about 3 % were Doctorates. The percentage of teachers with a Doctorate degree was higher in the conventional engineering discipline such as Mechanical engineering than in Computer science.

The comparison of results of DBMS and Thermodynamics workshops after scaling up the number of participants show that there was not much change in the demography of the participants due to scale up. However, we observed that as the number of remote centres were increased, the number of participants from rural colleges were also increased, thus, reducing the gap between the urban and rural college teachers. Participants' perception on the quality of instruction, quality of audio and video transmission, usefulness of workshop content were comparable for both small and large workshops. Comparison of overall satisfaction of the small and the large DBMS and Thermodynamics workshops show that the participants of the smaller workshop were more satisfied in comparison with the larger workshop. The mean difference between the two workshops was statistically significant at $p \leq 0.001$. Even though the outreach and inclusion were very good after scaling up, one of the drawbacks due to scale up was that due to paucity of time a large number of remote centres did not get an opportunity to interact live with the instructor at the main hub and that caused some dissatisfaction among the participants.

The survey showed that even though the participants attended these workshops in the distance mode, the overall experience of the participating teachers was very good. On an average about 90 % of the respondents rated their experience as excellent or good. Thus, effective use of ICT for teacher training can scale up the training activity significantly in the country and help in human capital formation. Such initiatives would not only help improve the quality of higher education, but also facilitate appropriate use of demographic dividend that India possesses. Thus, by using technology one can reduce the unequal access in teacher training in a large country like India.

Acknowledgments The authors wish to thank Prof. D.B. Phatak for allowing them to carry out the survey. The financial support for conducting the workshop was made available by a grant from the National Mission on Education through ICT (NMEICT), Ministry of Human Resource Development, Government of India, New Delhi. The e-learning software AVIEW, also supported by NMEICT, was provided by the Amrita Vishwa Vidyapeetham, Kollam, Kerala. An earlier version of this paper was presented in a workshop 'Harnessing Technology for Challenging Inequality' of Knowledge Forum held at the Tata Institute of Social Sciences, Mumbai on 7 March 2015. We are grateful to Professors. N.S. Siddharthan, Y.S. Rajan, S. Baskaran and Kannan Moudgalya for their useful comments and suggestions.

References

Adem, A. E. (2009). Teacher training through distance education: IICBA's experience. *Africa Education Review, 6*(1), 174–184.

AICTE. (2012). Annual report 2011–2012. http://www.aicte-india.org. Accessed November 30, 2014.

Arora, G. L., & Pandey, S. (1998). Teachers' continuing education: Shifting focus on distance mode. *Indian Journal of Open Learning, 7*(3), 255–265.

AVIEW. (2014). Amrita Virtual Interactive e-learning World—A multimedia e-learning platform. http://www.aview.in. Accessed October 12, 2014.

Berge, Z. L. (2007). Training in the corporate sector. In M. G. Moore (Ed.), *Handbook of distance education* (2nd ed., pp. 515–529). New Jersey: Lawrence Erlbaum Associates.

Bhattacharya, B. (2008). Engineering education in India—The role of ICT. *Innovations in Education and Teaching International, 45*(2), 93–101.

Davis, F. D. (1989). Perceived usefulness, perceived ease of use and user acceptance of information technology. *MIS Quarterly, 13*(3), 319–339.

Dixit, U. (2009). *The use of ICT in teacher training: Nepal's experience*. Paper Presented at 13th UNESCO-APEID International Conference and World Bank-KERIS High Level Seminar on ICT in Education, Hangzhou, People's Republic of China, November 15–17, 2009.

EDUSAT. (2013). Educational Satellite launched by ISRO. http://www.isro.gov.in/category-spacecraft/edusat. Accessed October 12, 2013.

IGNOU. (2014). Indira Gandhi National Open University. http://www.ignou.ac.in. Accessed November 10, 2014.

Kannan, K., & Narayanan, K. (2012). ICT-enabled scalable workshops for engineering college teachers in India. In R. Clothey, S. Austin-Li, & J. C. Weidman (Eds.), *Post-secondary education and technology—A global perspective on opportunities and obstacles to development* (pp. 127–146). New York: Palgrave Macmillan.

Kannan, K., & Narayanan, K. (2015a). A structural equation modelling approach for massive blended synchronous teacher training. *Journal of Educational Technology and Society, 18*(3), 1–15.

Kannan, K., & Narayanan, K. (2015b). Synchronous teacher training in India: A study of perceptions and satisfaction of the participants. *International Journal of Information and Education Technology, 5*(3), 200–207.

Lidstone, J., & Shield, P. (2010). Virtual reality or virtual real: Blended teaching and learning in a master's level research method class. In Y. Inoue (Ed.), *Cases on online and blended learning technologies in higher education: Concepts and practices, information science reference* (pp. 91–111). Hershey, PA: IGI Global.

Looi, C.-K., Hung, D., Bopry, J., & Koh, T.-S. (2004). Singapore's learning sciences lab: Seeking transformations in ICT enabled pedagogy. *Educational Technology Research and Development, 52*(4), 91–99.

NMEICT. (2013). National Mission on Education through ICT, Ministry of Human Resource Development, Government of India. http://www.sakshat.ac.in. Accessed October 12, 2013.

Norburg, A. (2012). Blended learning and new education logistics in Northern Sweden. In D. Oblinger (Ed.), *Game changers: Education and information technologies* (pp. 327–330). Boulder, CO: Educause.

NPTEL. (2014) National Programme for Technology Enhanced Learning. www.nptel.ac.in. Accessed November 30, 2014.

Pouezevara, S. L., & Khan, R. (2007). *Learning communities enabled by mobile technology: A case study of school-based, in-service secondary teacher training in rural Bangladesh*, Technical Report ADB TA NO. 6278-REG, Asian Development Bank.

Ramos, A. J., Nangit, G., Range, A. I., & Trinona, J. (2007). ICT-enabled distance education in community development in the Philippines. *Distance Education, 28*(2), 213–229.

Roseth, C., Akcaoglu, M., & Zellner, A. (2013). Blending synchronous face-to-face and computer supported cooperative learning in a hybrid doctoral seminar. *TechTrends, 57*(3), 54–59.

Sridhar, S. (2005). E-Government—A proactive participant for e-learning for higher education. *Journal of American Academy of Business, 7*(1), 258–268.

T10KT. (2014). Website for Train 10000 Teachers Workshop. http://www.it.iitb.ac.in/nmeict/home.html Accessed October 12, 2014.

Taylor, M. (2005). Access and support in the development of a visual language: Arts education and disabled students. *International Journal of Art and Design Education, 24*(3), 325–333.

Tinio, V. L. (2010). *ICT in Education*, Technical Report, UNDP. http://www.apdip.net/publication/iespprimes/eprimer-edu.pdf. Accessed March 3, 2012.

UGC. (2011). *Higher education in India, strategies and schemes during Eleventh Plan Period (2007–2012) for Universities and Colleges*, Technical report, University Grant Commission, New Delhi. www.ugc.ac.in/oldpdf/pub/he/heistategies.pdf. Accessed March 30, 2012.

UNESCO. (2001). *Teacher education through distance learning: Technology, curriculum, cost, evaluation*, Technical report, UNESCO.

Wang, Y., Chen, N.-S., & Levy, M. (2010). The design and implementation of a holistic training model for language teacher education in a cyber face-to-face learning environment. *Computers & Education, 55*(2), 777–788.

Wild, M. (2004). *Screen or page: Will the use of computer aided instruction improve phonological skills in year 1 classes?* Paper Presented at the British Educational Research Association Annual Conference, University of Manchester, September 16–18, 2004. http://www.leeds.ac.uk/educol/documents/00003882.htm. Accessed October 12, 2014.

Williams, P. (2005). Lessons from the future: ICT scenarios and the education of teachers. *Journal of Education for Teaching, 31*(4), 319–339.

Young, A., & Lewis, C. W. (2008). Teacher education programmes delivered at a distance: An examination of distance student perceptions. *Teaching and Teacher Education, 24*(3), 601–609.

Zhang, X., & Hung, S. (2007). Integrating of the high-tech and low-tech in distance teacher training in China: An insight from the case of Jiangsu Radio and Television University. *International Review of Research in Open and Distance Learning, 8*(1), 1–14.

MIX
Papier aus verantwortungsvollen Quellen
Paper from responsible sources
FSC® C105338

If you have any concerns about our products,
you can contact us on
ProductSafety@springernature.com

In case Publisher is established outside the EU,
the EU authorized representative is:
Springer Nature Customer Service Center GmbH
Europaplatz 3, 69115 Heidelberg, Germany

Printed by Libri Plureos GmbH
in Hamburg, Germany